SpringerBriefs in Ethics

Springer Briefs in Ethics envisions a series of short publications in areas such as business ethics, bioethics, science and engineering ethics, food and agricultural ethics, environmental ethics, human rights and the like. The intention is to present concise summaries of cutting-edge research and practical applications across a wide spectrum.

Springer Briefs in Ethics are seen as complementing monographs and journal articles with compact volumes of 50 to 125 pages, covering a wide range of content from professional to academic. Typical topics might include:

- Timely reports on state-of-the art analytical techniques
- A bridge between new research results, as published in journal articles, and a contextual literature review
- A snapshot of a hot or emerging topic
- In-depth case studies or clinical examples
- Presentations of core concepts that students must understand in order to make independent contributions

More information about this series at http://www.springer.com/series/10184

Vijayaprasad Gopichandran

Dynamics of Trust in Doctor-Patient Relationship in India

A Clinical, Social and Ethical Analysis

Springer

Vijayaprasad Gopichandran
Department of Community Medicine
ESIC Medical College and Postgraduate
Institute of Medical Science and Research
Chennai, Tamil Nadu, India

ISSN 2211-8101 ISSN 2211-811X (electronic)
SpringerBriefs in Ethics
ISBN 978-981-15-0345-0 ISBN 978-981-15-0346-7 (eBook)
https://doi.org/10.1007/978-981-15-0346-7

This Springer imprint is published by the registered company Springer Nature Singapore Pte Ltd.
The registered company address is: 152 Beach Road, #21-01/04 Gateway East, Singapore 189721, Singapore

This book is dedicated to Mrs. Bhuvaneswari and Mr. Gopichandran, for giving me life, nurture, values and freedom.

Preface

We live in troubled times with rapid erosion of trust in the physician-patient relationship. There is an increase in the reports of violence against physicians and healthcare providers. The crisis of mistrust is progressing and invading all aspects of the medical enterprise. In this context, there is a need to understand the dynamics of trust in the physician–patient relationship. What makes patients trust their physicians? What erodes the trust? Is trust a value in health care? Can trust be built through specific interventions? All these questions arise in this context. Much research on trust in the physician–patient relationship has emerged from the high-income-country settings. In the USA, the UK, Australia and many European countries, trust has been measured and quantified, and much research has been done on various aspects of trust in the physician–patient relationship. The literature on trust in the physician–patient relationship in India is sparse. There is a need to understand the trust dynamics in the Indian context.

This book is the output of doctoral research that I undertook between the years 2011 and 2015. The research was motivated by several questions that personally struck me as a practitioner of primary care in a rural clinic in south India. It started as a qualitative exploration and iteratively grew into a quantitative survey, scale development and validation exercise. Several peer-reviewed scientific publications have emerged from this Ph.D. research. These are independent stand-alone research publications, and they are richly cited throughout this book. However, none of these publications provide a comprehensive overview of the physician–patient trust dynamics that emerged from the overall study. I felt the need to write this book, to give the broad bird's-eye view of the trust dynamics. I hope this book fulfils that purpose.

The chapters of this book have been restructured and organized differently compared to the Ph.D. dissertation. The chapters describe the various phases of the research work, and I have strung them together to reflect the sequence in which the understanding of the phenomenon of trust unravelled itself to me. I have consciously removed some sections from the main dissertation, for the purpose of brevity and clarity. Some chapters carry statistical analysis, numbers and tables and their interpretations as they are found in the dissertation. These help to make sense of the scientific rigour that went into the data analysis and interpretation.

I have written this book in an academic style and for readers from medicine, health care, medical sociology, health service sector, health system research, medical anthropology and other specialties that deal with physician–patient relationship issues. Though it is written in an academic style, even the lay reader can make sense from many chapters of the book. The scale development section of the book may be of interest to those working in the area of psychometrics and patient-reported outcome measurement.

I do not claim that the theory of trust I propose in this book is the final and accurate theory of trust in physicians in a developing healthcare setting like India. The area of trust dynamics needs more research and deeper understanding. However, I think this book will contribute to widening the scope of trust research in India and will expand the various dimensions of trust as we understand it today. I wish all the readers a productive reading experience.

Chennai, India Vijayaprasad Gopichandran

Acknowledgements

I sincerely acknowledge and thank Prof. Satish Kumar Chetlapalli, my research guide and mentor, whose constant encouragement and support made this research possible. In the long and tedious journey of Ph.D. research, a good mentor like Prof. Satish is a blessing and I am forever thankful to God for giving him as my guide. My teacher, mentor, friend, philosopher and guide, Dr. K. P. Misra, Senior Consultant Cardiologist, who moulded and strongly influenced my thinking in the young formative years, is the guiding force behind this book. Sir, you are not with us today, but I am sure you will be extremely proud to see this book, and highly critical of the mistakes in it! Various teachers before, during and after my medical school—Mrs. Jamuna Mohan (school), Mrs. Alamelu Chidambaram (school), Mrs. Pramila Theodore (higher secondary school), Prof. N. Karunanidhi (general medicine), Prof. D. Bennett (general medicine), Prof. Deepak Kabir (general surgery), Prof. Pari (general surgery), Dr. Usha Sriram (endocrinology), Dr. Arjun Rajagopalan (general surgery), Prof. J. P. Muliyil, Prof. K. R. John, Prof. Jasmin Helan Prasad—have left indelible impressions in my life and work. My colleagues and mentors, Dr. Amar Jesani and Prof. Mala Ramanathan, have both been sounding boards for my research and helped me understand my research better. They have both given me platforms to showcase my work and encouraged me beyond my wildest dreams. I am also sincerely thankful to Prof. Angus Dawson, for reviewing my dissertation and giving me critical inputs on it. I always look up to Prof. Dawson with respect and admiration for his work in public health ethics. I also must thank Ms. Geetha Veliah and Ms. Meggha Sheth, two of my dearest friends, who have encouraged me to complete my Ph.D. as well as write this book. Lastly, I thank all my friends and family for their support and encouragement.

Contents

About the Author

Dr. Vijayaprasad Gopichandran currently works at the Department of Community Medicine, ESIC Medical College and Postgraduate Institute of Medical Science and Research, KK Nagar, Chennai. His areas of interest include non-communicable disease epidemiology, health communication, doctor–patient relationships, and medical and public health ethics. He serves as an editor for the Indian Journal of Medical Ethics since 2014. He is also a consultant for the World Health Organization for developing ethical guidelines for public health surveillance and developing a curriculum for training in ethics of implementation research. He has more than 17 years of primary care clinical work experience in outpatient and inpatient settings and 9 years of clinical experience in community outreach projects. He has been actively involved in numerous community-based epidemiological and bioethics research projects, leading to over 50 publications in peer-reviewed national and international journals.

Chapter 1
Introduction

Abstract Trust is the optimistic acceptance of one's vulnerability in the best belief that the trusted party will do whatever is in the best interest of the one who is trusting. There is substantial scholarship of trust in health care, with specific focus on trust in the physician–patient relationship. From existing literature, there is an understanding of trust in physicians as comprised of competence, fidelity to the patient's welfare, honesty and confidentiality. There is also substantial research and literature on factors influencing trust in physicians. Despite this sound understanding of the dimensions and determinants of trust in physicians, studies show that interventions to build trust in physicians do not work. Most of this research on trust emerges from developed healthcare settings in the West. There is a need for understanding trust in physicians in developing healthcare settings. This chapter describes the existing literature on trust in physicians from the Western context and justifies the need for research from developing healthcare context.

Keywords Trust in physicians · Developing healthcare settings · Dimensions · Determinants · Influencing trust

Robert Plutchik, the renowned psychologist, has defined eight basic emotions which are biologically primitive and necessary for reproductive fitness of animals including humans (Plutchik, 2003). He classifies trust as one among the eight primitive human emotions. Trust has been identified as the fundamental basis of all relationships within and between social groups. Trust is a measure of belief in honesty, fairness and benevolence of another person. In a human relationship, loss of trust will not happen if the transgression is perceived as a failure of competence, but trust will be broken if the perception is a transgression of honesty, fairness or benevolence (Mechanic, 1996b).

Health care is a dynamic social institution, and trust forms an integral part of human interactions with the healthcare system. Trust in physicians is important given the state of vulnerability inherent in illness. Patient trust in the physician has been defined as a collection of expectations that the patients have from their physician (Anderson & Dedrick, 1990). These expectations could include a detailed examination, a correct diagnosis, appropriate tests, appropriate treatments, communication and finally a positive health outcome. Another interesting definition of trust, which is

V. Gopichandran, *Dynamics of Trust in Doctor-Patient Relationship in India*, SpringerBriefs in Ethics, https://doi.org/10.1007/978-981-15-0346-7_1

apt for the healthcare setting, is 'an unwritten agreement between two or more parties for each party to perform a set of agreed-upon activities without fear of change from any party' (Shore, 2003).

Further, healthcare trust can be manifested as trusting attitudes and trusting behaviours. These are not necessarily the same. While it is possible that a person may exhibit trusting behaviour by seeking health care from a physician, the patient may not necessarily have a positive trusting attitude towards her (Hall, Dugan, Zheng, & Mishra, 2001). Trust in health care has been shown to lead to several important outcomes. Greater trust in physicians leads to greater adherence to treatment, greater follow-up, lesser need for second opinions, lesser need for investigations, reduction in healthcare cost and a perceived placebo effect of treatment.

Research on trust in physicians from the developed world context is available in plenty. Most studies have emerged from the USA and the UK and some studies from Australia. However, it is important to note that trust in physicians in these settings is likely to be very different from trust in physicians in developing countries like India. In developing healthcare settings where there is a limitation of resources, lack of universal health coverage, low public expenditure on health care, high out of pocket expenditure on health and poorly regulated private health care, the dimensions and determinants of trust are likely to be different. Economic factors, uncertainties in access to health, emotional factors and implicit acceptance of paternalism in clinical practice are all likely to operate in these healthcare settings, thus leading to different dimensions of trust. A sound understanding of trust in physicians is important for proper assessment of the social dimension of healthcare delivery in these settings.

Dimensions of Trust in Physicians

Trust in physicians has been defined by five key dimensions by Hall et al., in their review (Hall et al., 2001). These dimensions are fidelity, honesty, competence, confidentiality and global trust. *Fidelity* means the physician takes a genuine interest in the patient, does all that is possible to the best of his/her ability to help the patients and does not exploit the vulnerability of the patient. Caring, respect, advocacy and avoiding conflicts of interest are the various subdomains in this dimension of trust. *Competence* means avoiding mistakes and working towards achieving optimal health for the patient. Often technical competence is not assessed by the patients due to the inherent difference in the level of knowledge between the physicians and the patient. But communication competence is usually assessed. Communication skills which relate to competent medical decisions, such as taking a detailed history and making the patient feel comfortable enough to reveal sensitive personal information, are important for building trust. Also important are making eye contact, being polite to the patient, expressing genuine concern by appropriate body language, etc., which can influence trust. *Honesty* as a dimension of trust can have two sides to it. On one hand, honesty in admitting to mistakes, transparency in healthcare decision making and full disclosure of all health-related conditions to the patient can help in increasing trust. But one concern is that honesty in disclosing lack of knowledge or disclosing a conflict of interest may lead to erosion of trust. *Confidentiality* is another important expectation and dimension of trust. But there is scope for exploring confidentiality, the extent of confidentiality expected by patients and specific conditions where confidentiality is emphasized by patients. Several authors have studied confidentiality as a dimension of trust and have come out with inconclusive results. The expectation of confidentiality may depend largely on the type of illness, severity of illness and socio-cultural characteristics of the patients. Therefore, confidentiality as a dimension of trust needs greater research, especially in developing country settings. Finally, *global trust* is the more holistic dimension of trust which cannot be captured by any of the other dimensions. It can also be viewed as the basket term to refer to other dimensions that contribute to trust.

Mechanic and Meyer in their paper argue that though patients can differentiate between the behaviorial and technical dimensions of trust, it is not easy for them to point out what makes them trust the physicians. They do not know what exact dimension of trust is more important than others (Mechanic & Meyer, 2000). This may be particularly true in the setting of developing countries where the level of awareness about health care and participation in health care is significantly lesser than the developed settings.

From another perspective, the trust may have two major dimensions to it—confidence and faith. Confidence is an objective assessment of the physician's performance based on experience or based on word of mouth from other members of the community. Thus, trust-as-confidence is an objective assessment which is done consciously by the patients. Faith is a belief in the physician which stems from an emotional plane. Some scholars define faith as an irrational belief not based on objective assessments. Trust-as-faith is blind (Gatter et al., 2002). Another aspect of trust that needs to be understood is that trust and distrust are not two sides of the same coin. In fact, a patient could have both trust and distrust at the same time. This theoretical concept was first described by Lewicki et al. They propose a matrix in which the four quadrants represent low trust–low distrust, low trust–high distrust, high trust–high distrust and high trust–low distrust. The low trust–low distrust segment is usually the one in which most human relationships start. Then, they slowly progress to one of the other three segments based on experiences. In health care, the most desirable is the fourth quadrant with high trust and low distrust. But most healthcare trust is in the high trust–high distrust quadrant, which is also referred to as the 'trust but verify' quadrant. Patients place high trust in the physician, but they prefer to verify by means of second opinions (Lewicki, Mcallister, & Bies, 2008).

There is strong evidence from qualitative explorations that trust in physicians is a state and not a trait. Respondents in qualitative studies were able to recall specific incidents and anecdotes which increased or reduced their trust in the physician. Also, there was little correlation between trust in a physician and general trust in people and general trust in the health system. This again indicates that trust is specific to individuals and situations. Trust was also found to be different in different physicians for the same patient. Thus, trust is largely a 'state' which depends on the experiences.

It is dynamic and keeps changing with experiences (Hall, Zheng, et al., 2002; Thom, Hall, & Pawlson, 2004). This trust which is characterized here is of great value in all human relationships, especially in healthcare relations.

Trust in health care can be broadly viewed from two important perspectives: trust in physicians, nurses and other providers which is interpersonal trust and trust in the system which is social trust or institutional trust. These two types of trust are closely related to each other. One might strongly influence the other. Interpersonal trust in the physician may give a positive trusting attitude towards the system or institution that the physician is affiliated to and vice versa (Hall et al., 2001). In developed countries, there have been reports that the general level of trust of the public in the health system is declining. Trust in the public health system in the USA has been shown to have declined over the past 30 years (Mechanic, 1996a, 1998). Similarly, there has been reported a reduction in trust levels on the National Health System (NHS) in the UK (Mechanic, 1996a). But despite such overall erosion of trust in the system, there remains a great amount of trust in physicians. Given the extreme complexity and fragmentation of the modern social structure, institutional trust plays a crucial role as it involves many social interactions at a time. Thus, institutional trust acts as an assurance for interpersonal trust in modern society (Misztal, 2013). Professional and ethical codes within the health systems help to strengthen the trusting relationships between patients and the system (Mechanic, 1996a). Funding mechanism within the system strongly influences public trust. Appropriate funding allocation, rationing, solidarity in fund distribution for health care and procedural justice and fairness in allocation have all shown to enhance public trust in health care (Mechanic, 2001).

Van der Schee and colleagues present an interesting conceptual framework to explain how interpersonal trust and institutional trust are interrelated. Strong healthcare systems provide good quality care and provide institutional entitlements or guarantees. These are delivered by the providers in the system. The patients encounter the providers and their interpersonal trust builds. This interpersonal trust is influenced by the system's entitlements and quality of care. Moreover, the patients also develop trust in the system through media image and network communications (Van der Schee, Braun, Calnan, Schnee, & Groenewegen, 2007).

Expectant trust is the trust that the patient has on the physician during the first clinical interaction, whereas experiential trust builds over time as a result of positive interactions with the physician (Goold, 2002). If the patients have negative experiences with the physician, especially in the aspects of communication, it leads to a reduction in trust (Keating et al., 1997). Several studies have reported that the provision of full information about the illness and active engagement of the patient in clinical decision making leads to increased trust (Schneider & Schneider, 1998). These are forms of experiential trust. Expectant trust largely depends on general trusting tendencies of the patients, second-hand information from other patients and certain indicators of trustworthiness that the patient perceives before the clinical interaction. For example, the patient may develop a trusting attitude by seeing a huge crowd waiting outside the clinic to see the physician when compared to an empty waiting room.

Value of Trust in Physicians

Trust in physicians has both instrumental and inherent value. Based on this, trust is further classified as a strategic trust which is of instrumental value and moralistic or altruistic trust which is of intrinsic value. In the strategic sense, trust is intentionally exposing one's vulnerability and taking the extra risk of being exploited by the physician so that one benefits out of the patient–physician relationship. In the altruistic sense, trust is a belief in the physician's intent for larger good as a gift to the community and a belief that the profession is noble and therefore the professionals are also of a high ethical standard. Therefore, strategic trust is a calculated cognitive phenomenon, whereas altruistic trust is an affective emotional phenomenon (Gilson, 2003). Zaner and colleagues believe that trust in physicians is not optional. According to them, trust is an integral and implicit component of all healthcare interactions (Zaner, 1991). Others propose that fiduciary duties be assigned to physicians to protect this inherent trust in the physician–patient relationship (Illingworth, 2002).

Studies have shown that higher trust in physicians translates to better adherence to treatment. Thom et al. in their elegant study on trust in physicians showed that 62% of patients who are in the highest quartile of trust in their physician reported complete drug adherence, whereas only 14% in the lowest quartile of trust did so (Thom et al., 2004). Similarly, patients who had high trust in their physicians were more likely to follow good self-management behaviours including smoking cessation and safe sexual practices (Safran et al., 1998). Trachtenberg et al. showed that higher levels of trust in physician were strongly associated with patients' active involvement in their treatment as indicated by wanting to be more in control over their treatment and seeking health care for most problems (Trachtenberg, Dugan, & Hall, 2005). In a cross-sectional study done in Michigan, it was found that patients with higher out of pocket expenses were more likely to stop their medications because of the expenses when physician trust was low. Thus, trusting patient–physician relationship may reduce the impact of high cost on medication adherence (Piette, Heisler, Krein, & Kerr, 2005). It was also found in the study by Thom et al. that patients in the highest quartile of trust had a higher chance of following up and continuing care with the same physician compared to those in the lowest quartile (Thom, Ribisl, Stewart, Luke, & Physicians, 1999).

A strong trusting relationship between the physician and patient leads to patients' perception that the treatment proposed by their physician will have good outcomes. Therefore, they have positive expectations from the treatment and hence adherence is greater. Strong patient–physician relationships promote self-efficacy among patients (Ohya et al., 2001). There are many mechanisms of how a trusting relationship improves self-efficacy. Mastery over the behaviour in trusting settings leads to positive reinforcement. A trusted physician's encouragement and motivation play an important role in promoting self-efficacy. Moreover, lower levels of anxiety and self-doubt that accompanies the confidence in a trusted physician can also promote self-efficacy. Using rigorous structural equation models, Lee et al. showed that patients who trusted their physicians had better self-efficacy and positive outcome expectations which in turn led to better self-rated health and objective outcomes (Y.-Y. Lee & Lin, 2009). In a study of diabetic patients at Wake Forest University, it was seen that higher trust of the patients on the physician led to fewer difficulties in following self-management behaviours (Y. Lee & Lin, 2009).

In the treatment of certain diseases, there is a need to collect and record certain sensitive information such as sexual behaviours from patients. A trusting relationship between the physician and patient helps the patient to disclose sensitive information. This, in turn, leads to appropriate treatment decisions (Thom et al., 2004). A higher trust may lead to reduced need for the second opinion about treatment decisions and reduced need for unnecessary tests and investigations. This may reduce the unnecessary cost of health care (Thom et al., 2004). In a study, it was shown that higher trust led to a significantly higher improvement in self-reported health status two weeks after a visit to the physician after adjusting for multiple potential confounders such as patient characteristics, physician characteristics and the characteristics of the medical visit (Thom, Kravitz, Bell, Krupat, & Azari, 2002). A population-based study in Sweden showed that a low level of trust in the healthcare system and low levels of general trust led to poor self-rated health. This was attributed to lower treatment-seeking behaviour during the illness which was associated with low levels of trust (Mohseni & Lindstrom, 2007). Thus, trust is not only of intrinsic value but is also instrumental in these aspects. Therefore, measuring and promoting trust is an important exercise in its own merit.

Studies of trust in physicians among African Americans in the USA have shown that low levels of trust lead to lower utilization of healthcare services, preventive services and surgical treatments (LaVeist, Nickerson, & Bowie, 2000). In a study, it was shown that trust in the physician was the strongest predictor for the initiation of antiretroviral therapy among African Americans and Hispanic Americans living with HIV/AIDS (Altice, Mostashari, & Friedland, 2001). One study showed that a lack of trust in the health system was associated with significant psychological distress (Ahnquist, Sarah, & Lindstro, 2010).

Factors that Influence Trust in Physicians

In some studies, the age of the patient has been shown to have an influence on trust. Older the age, greater the trust in the physician (Hall et al., 2001). Trust builds iteratively. Therefore, longer the duration of a relationship, greater is the trust (Kao, Green, Davis, Koplan, & Cleary, n.d.). This is more relevant in the settings where patients have long-term relationships with providers, such as the National Health System of the UK. But in settings like the public health system in developing countries, long-term interpersonal relationships are rare because of rotation of physicians within the system, and hence, this association between age of the patient and trust is less likely.

Among women seeking obstetric care, their socio-economic status had a big bearing on the trust. Women from the middle-class social background who had a high perception of risk of obstetric outcomes preferred private obstetricians even if they had to pay more (Riewpaiboon, Chuengsatiansup, Gilson, & Tangcharoensathien, 2005). The greater vulnerability offered by certain social conditions may tilt the trust dynamic, either way, the vulnerability may increase blind trust, or it may make a person cautious and less trusting. Such trust dynamics are important to understand.

In an interesting analysis, the researcher found that religious affiliation and level of religious activity positively correlated with trust. Those who were very active religiously tend to have higher trust in physician and in health care in general. One explanation for this association could be that religious activity has an influence on the general trust of the individual which in turn influences the trust in physicians and health care (Halbert, Armstrong, Gandy, & Shaker, 2006). Among cardiac patients, it was seen that lesser education and greater perceived control over their illness led to greater patient trust in physicians (Benjamins, 2006). Patients with certain illnesses such as congestive heart failure and depression reported greater levels of trust (Kayaniyil et al., 2009).

Some important characteristics of the physician such as their attitudes, approach and behaviour have an important bearing on trust. Researchers have recorded the interactions between standardized patients and primary care physicians and noted the level of trust that patients attending these physicians placed on them. On multivariate analysis of factors influencing the trust, certain patient-centred behaviours such as exploring patients' disease and illness experience and greater duration of the visit were positively correlated with trust. Greatest trust was reported by patients visiting family physicians (Fiscella et al., 2004).

Experiences related to the interpersonal aspect of physician behaviour such as giving enough time to the patient to explain the reason for the visit, providing answers to patient's questions in an understandable manner, taking enough time to answer patient's questions, enquiring about the effect of family situation on health, involving the patients in decision making and providing as much medical information as the patient wants were determinants of both trust in physician as well as satisfaction (Keating et al., 1997). The physician behaviour is a direct reflection on the genuine intentions and the covenant of care between the physician and patient. In rural Cambodia, trust in the physician was rated as the second most important reason to choose a private provider, whereas it was the fifth reason to choose a public care provider. While public providers were trusted for their skills, abilities and effective referral system, the private providers were trusted for friendliness, easy approachability and accessibility and being careful and thorough (Thom et al., 2002).

Low patient trust was associated with complaints that a requested service was not provided by the physician. It was also seen that the level of trust was positively correlated with satisfaction with care, reporting of a positive outcome of treatment and desire to follow the physician's advice (Ozawa & Walker, 2011). Interestingly, the physician's attire seems to influence trust. It was observed that patients preferred the professional attire with a white coat, followed by scrub suits, business dress and finally casual clothes. They felt that they could talk more openly about their social, sexual and psychological problem with a physician in professional attire. There was also a gender dimension to this preference with dress code reported as being more important for female physicians compared to male physicians (Rehman, Nietert, Cope, & Kilpatrick, 2005). Physician's perceived technical competence, honesty and patient-centred behaviour were all strongly correlated with patients' trust in physicians (Hillen, De Haes, & Smets, 2011). Certain patient-friendly behaviours such as being comforting and caring, demonstrating competency, encouraging and answering questions and explaining were associated with trust. The behaviours least important for the trust were gentleness during the examination, discussing options/asking opinions, looking in the eye and treating as an equal (Thom, 2001).

A study from Sri Lanka showed that treatment-seeking behaviour depended on public trust which is determined by the public perception of functioning of the health system. In Sri Lanka, despite the strong public health system, poor quality of interpersonal relationships prevents even the poor from accessing the system. Thus, deteriorating interpersonal relationships between the physicians and patients could act as potential threats to universal health access even in robust public health systems (Russell, 2005). Transparency is a very important determinant of trust. The World Health Organization guidance document on Pandemic Influenza Preparedness and Response emphasizes the need for commitment to transparency and credible actions. Public perceptions will play an important role in trust-building. Trust is very important in times of pandemics in order to curtail the spread of the disease (Larson & Heymann, 2010).

Reporting of unmet needs for health care and delayed care is higher among patients who have lower levels of fiduciary trust in the physician. This negative association seems to be more pronounced among the poor, racial minorities and those who are not covered by a health insurance scheme. This association between trust and unmet and delayed health care could be reciprocal. A poor trust could lead to delayed or no access to health care and poor access to health care could lead to poor trust (Mollborn, Stepanikova, & Cook, 2005). In certain communities, gatekeeping of health care is done as a step to reduce healthcare cost by streamlining referrals, reducing unnecessary investigations and procedures. Communities which had greater gatekeeping activity had lesser trust in health care compared to communities which had lesser gatekeeping (Haas, Phillips, Baker, Sonneborn, & McCulloch, 2003). Continuity of care was found to be significantly associated with patients' trust in their physician. This was reported in a study from two countries, the UK and the USA (Mainous, Baker, Love, Gray, & Gill, 2001). Patients who had a regular physician, to whom they could go, had a higher level of trust compared to those who did not. In another study, practice climate, defined as supportive interactions between practitioners and staff in the clinics, led to significantly higher trust among the patients. The authors substantiate that supportive and interactive positive relationship in the practice environment leads to a reduction in the power differential between the physician and patient thus leading to higher levels of trust (Becker & Roblin, 2008).

The level of trust also varied based on the difference in social capital in the communities. Social capital ensures reciprocity and general trust in people. This leads to greater levels of trust in the physician. This social capital of the community is a significant determinant of trust in health care in that community (Ahern & Hendryx, 2003). Trust that patients had on their physicians depended on having a choice of physicians, having a long-term relationship with the physician and higher levels of trust in the managed care organization. Patients believed that paying the physician per test or procedure rather than a fixed monthly amount would not adversely affect their care. They also believed that paying the physicians more for ordering fewer tests would make the quality of their care worse. They were also willing to accept higher co-payments for obtaining necessary tests (Kao et al., n.d.). Households, where the burden of healthcare expenditure is high, had lower trust in the physician and also had a negative assessment of the quality of care. Thus, rising healthcare cost is an important determinant of patient trust even in a developed setting (Cunningham, 2009).

Conflicts of interest played an important role in patient–physician trust. Patients who believe that their physicians receive gifts from the pharmaceutical industry had lower levels of trust in their physicians compared to those who did not believe so. These patients also reported a high level of health system distrust. The patients who believed that almost all physicians receive some gift from the pharmaceutical industry reported even higher levels of distrust in the health system (Grande, Shea, & Armstrong, 2012).

The above factors seem to indicate that interventions could be devised to appropriately address the physician, patient and health system characteristics which in turn can influence trust in health care. For any meaningful interventions for influencing trust in health care, there is a need for effective measurement of the construct.

Building Trust in Physicians

Three studies have looked at the impact of trust-building interventions on the patients' trust in their physicians. Thom et al. gave training to physicians on certain behaviours which were associated with increased patient trust. The training focused on clearly addressing the patients' concerns, thorough history taking and physical examination, involving the patient in decision making, addressing the doubts and questions that the patients have and arranging for follow-up care. There was only a two-point difference in the post-trial physician trust scores between the intervention and control groups. There was no statistically significant difference between the two groups (Thom, 2000).

Hall et al. studied whether disclosure of the information about the incentives that the physician received for complying with managed care protocols increased the trust of patients on their physicians. He showed that there was a marginal 1.4% increase in the patient trust after the disclosure of the incentives. It was further reported that the increase was only among those who had a baseline high level of trust and not among those who had low trust at baseline. The authors also report that the disclosure had a greater impact on trust in the insurer rather than trust in the physician (Hall, Dugan, Balkrishnan, & Bradley, 2002).

Thompson et al. studied the impact of three interventions compared to control on the trust of the patients on physicians. When the patients enrolled in an HMO, three different types of induction visits were offered. The first was a physician-only visit, the second was a physician visit followed by a visit by a health educator, and the last one was group visit of eight new members led by a physician and a health educator. In the group visit, the trust on physician was higher compared to control, but this was not seen in the other two groups. Overall trust in a health plan was higher in the 'physician + health educator' and the group visit groups (Thompson, Gee, Larson, Kotz, & Northrop, 2001). A systematic review of interventions to build and promote trust in physicians and the health plan revealed that there is insufficient evidence to show that there is any intervention that can promote trust in physicians (Mckinstry, Ashcroft, Car, Freeman, & Sheikh, 2008).

The futility of the exercises to build trust implies that the understanding of trust and factors influencing it is incomplete. All the studies of trust that have been discussed until this point have been reported in the developed country settings such as the USA, UK and Australia. There is very little understanding of any trust in health care in developing country settings.

Trust in Physicians in Developing Healthcare Settings

There are several unique features of the healthcare systems in developing countries.

- Many developing countries have strong inequities in health indicators, health status, health access and social determinants of health (Makinen et al., 2000).
- Universal health access is not available in many of these countries. The models of health care in Brazil and Thailand which have made great strides towards the achievement of universal health access are interesting, but many countries do not have such robust primary healthcare systems.
- Budget allocation to the health sector is very low in many developing countries, thus leading to higher private expenditure in health care.
- The poorly organized public health sector in many developing countries has led to the burgeoning of the private sector, which in some situations is profit-driven, inequitable and unregulated.
- Quality of health care leaves much to be desired in many developing countries.

Trust in physicians in this setting has not been explored in great depth so far. A recent study from Pune, a metropolitan city from the western part of India, has explored trust in health care in the hospital setting. They reported that about 60% of the patients had a high level of trust in their physician and the remaining 40% had some amount of reservations about their physicians. Further, they reported that men had higher trust in physicians compared to women, but no other factors were significantly associated with trust. Physician–patient gender concordance led to higher levels of trust. Further higher trust led to better patient enablement for self-management (Banerjee & Sanyal, 2012). One of the major limitations of this study is that it used the Trust in Physician Scale developed in 1990 by Anderson and Dedrick, which was validated in the USA. The dimensions and determinants of trust in the setting of developing countries are likely to be different from that in the USA. The dimensional structure of the responses was not assessed or reported in this study. This points to a need for a thorough understanding of trust dynamics in these healthcare settings. This being one side of the problem, the other side is that even in settings where there is some data on the dimensions and determinants of trust, trust in health care seems to be a highly evolving and dynamic construct.

There has been a rapid growth in technology in the medical field starting from the late twentieth century. This has resulted in major advances in the understanding of disease processes, and in the diagnosis, treatment and course of diseases (Cutler & McClellan, 2001). The evidence-based medicine movement which evolved over the past four decades is accepted as the *sine qua non* of good quality medical care (Straus & McAlister, 2000). Alongside this development is the growth of business models in health care. Corporatization of medical care has given an impetus to the advancement of technology in medicine in India (Jindal, 1998). Health care of high standards has reached many low- and middle-income countries. Yet, large segments of the population still do not have access to even basic health care. Many of these countries are struggling to establish universal health coverage. The development of the human rights movement post-World War II, and the more recent discourse on realizing health as a human right, has significantly contributed to the understanding of these health inequities and the need for universal health coverage in these countries. At the same time, the era of information technology, heralded by the development of the Internet, and communication technologies like mobile phone services has significantly shrunk the world. Communities are being defined differently, with less personal and more virtual interactions. People seeking health care may also visit the Internet for information. These developments have affected people's trust in health care in various ways. It has been suggested that developments in the socio-political scenario of healthcare have affected people's trust in healthcare providers (Pearson & Raeke, 2000).

From historical times, the physician–patient relationship has been given a special status, and several measures such as various codes of conduct from the Hippocratic oath to modern biomedical codes of ethics have been adopted to protect its sanctity. Trust in the physician and health care was unquestioned and implicit, based on an expectation that the physician followed professional ethics. This trust in health care and providers changed with growing evidence that physicians sometimes acted against their patients' interests. The Nuremberg trials post-World War II provided enough grounds for mistrust in the profession. As a result, greater emphasis was placed on autonomy and self-determination of individuals, by the international organizations which emerged to codify the conduct of physicians. It was emphasized that patients should be made equal partners in medical decision making. Some studies found that when patients actively participated in the decision making, the outcomes were better (Fraenkel & McGraw, 2007). In this period, it became an ethical requirement for physicians and researchers to obtain voluntary informed consent for participation in medical research as well as for treatment and procedures.

As patients become increasingly able to make informed decisions, their trust in the physician may reduce as they start suspecting that the physicians might act in their own self-interest rather than the best interest of the patient (O'Neill, 2004). Patients, who have blind trust in their physician, tend to participate less in the clinical interaction and allow physicians to make most decisions. This can have negative consequences. The physician may fail to do the best in the circumstances as they may not feel accountable for their actions. There is also the potential for exploitation as physicians know that their decisions will not be questioned. Thus, with the rise in importance of autonomy in the patient–physician interaction, patients' trust may no longer be 'blind trust'. Autonomous patients may choose to have 'calculated trust' in which they weigh their choices and make a calculated choice to trust the physician. This phenomenon may be rising in developing healthcare settings.

The latter part of the twentieth century heralded a rapid growth in information technology. While the Internet is a powerful source of information, the ease of access to information can be a double-edged sword. There is a proliferation of information, without regulation to ensure its accuracy, and some of it can be misleading. The ease of access to information has empowered patients to ask questions but may also have led to a decrease in trust. When patients have unlimited access to information, their trust in the physician should be understood as a dynamic phenomenon. Patients expect that the physician will do what is best for them. But they are also alert enough to ask questions and verify the physician's advice if need be. This relationship has been described as 'trust but verify' (Y.-Y. Lee & Lin, 2009).

The growth of the private sector in health care in these developing healthcare settings has led to significant improvements in the quality of health care and technology but has also caused glaring disparities in health. The business model of health care has led to scepticism among patients. Patients who pay large sums of money out of their pocket may wonder if their physicians are doing all that is best for their treatment. They may wonder whether their physicians' decisions are motivated by money. At the same time, there is some evidence that in developing countries, trust in private physicians is higher than in the public health system. They may believe that paying money buys them trustworthiness in the private health system. This can be described as a 'sceptical trust', the patient's strategic acceptance of his vulnerability while remaining sceptical in the belief that money determines trustworthiness.

Another important development which has a significant impact on patient–physician trust is the growth of evidence-based medicine (EBM). The evidence-based medicine movement has been criticized for disrespecting the value of clinical experience and expertise. Its protocol-driven practices ignore patient preferences and can increase healthcare costs. However, the presence of standard treatment protocols which are publicly available may also lead to increased accountability. By demystifying medical treatment, EBM also leads to a reduction in a blind trust in the physician. The patient gathers information about the disease and treatments, and the trust he/she develops is more in the standardized protocols and treatment procedures than in the physician. This makes the trust impersonal.

The model of trust in health care has evolved significantly, as described in the previous paragraphs. The models presented here only describe the extent to which the trust is blind or well informed. Trust is a dynamic construct. The dynamism referred to here is the responsiveness of the nature of trust in physicians to the social situation. Assessment of trends of trust in physicians over time should take into consideration the interaction of variables such as the social environment, political circumstances and scientific developments. It is more likely that the form of trust has changed, and not trust itself. A true loss of trust in health care is reflected as a reduction in trusting behaviours. A typical example is an adverse event following immunization for measles in Tamil Nadu. Following the death of four children after administration of the measles vaccine, the measles vaccine coverage rate reduced drastically. This is a direct reflection of the loss of trust in the measles vaccine and the system. But reports of reduced trust in the public health system must be assessed with care as they are not associated with concurrent distrustful behaviour.

In summary, trust in physicians is an optimistic acceptance of the vulnerability of the patient in the belief that the physician will do what is in the patient's best interest. Trust can be interpersonal between the physician, nurse, allied health professional, and the patient or it could be institutional. Trust is an abstract construct which has the dimensions of competence, honesty, confidentiality, fidelity and global trust. Trust in health care can be influenced by a wide variety of physician, patient and health system factors. Trust has important beneficial effects in the clinical encounter. Trust increases adherence to treatment, follow-up, revealing of sensitive health information, a perceived placebo effect, reduced need for second opinions and reduction in healthcare costs. Though the determinants of trust are understood well, it has not been possible to demonstrate any effective intervention to promote trust in health care. Trust in health care is an evolving phenomenon from 'blind faith' to strategic, calculative, sceptical and objective trust. This background understanding of trust in physicians is essential to delve into the exploration of trust dynamics in the physician–patient relationship in developing healthcare settings.

References

Ahern, M. M., & Hendryx, M. S. (2003). Social capital and trust in providers. *Social Science & Medicine, 57*, 1195–1203. https://doi.org/10.1016/S0277-9536(02)00494-X.

Ahnquist, J., Sarah, P., & Lindstro, M. (2010). What has trust in the health-care system got to do with psychological distress? Analyses from the national Swedish survey of public health. *International Journal for Quality in Health Care, 22*(4), 250–258.

Altice, F. L., Mostashari, F., & Friedland, G. H. (2001). Trust and the acceptance of and adherence to antiretroviral therapy. *Journal of Acquired Immune Deficiency Syndromes (1999), 28*(1), 47–58.

Anderson, L. A., & Dedrick, R. F. (1990). Development of the trust in physician scale: A measure to assess interpersonal trust in patient-physician relationships. *Psychological Reports, 67*(3_suppl), 1091–1100.

Banerjee, A., & Sanyal, D. (2012). Dynamics of doctor-patient relationship: A cross-sectional study on concordance, trust, and patient enablement. *Journal of Family and Community Medicine, 19*(1), 12.

Becker, E. R., & Roblin, D. W. (2008). Translating primary care practice climate into patient activation: The role of patient trust in physician. *Medical Care, 46*(8), 795–805.

Benjamins, M. R. (2006). Religious influences on trust in physicians and the health care system. *The International Journal of Psychiatry in Medicine, 36*(1), 69–83.

Cunningham, P. J. (2009). High medical cost burdens, patient trust, and perceived quality of care. *Journal of General Internal Medicine, 24*(3), 415–420.

Cutler, D. M., & McClellan, M. (2001). Is technological change in medicine worth it? *Health Affairs, 20*(5), 11–29.

Fiscella, K., Meldrum, S., Franks, P., Shields, C. G., Duberstein, P., Mcdaniel, S. H., et al. (2004). Patient trust: Is it related to patient-centered behavior of primary care physicians? *Medical Care, 42*(11), 1049–1055.

Fraenkel, L., & McGraw, S. (2007). Participation in medical decision making: The patients' perspective. *Medical Decision Making, 27*(5), 533–538.

Gatter, R., Goldner, A., Greaney, T. L., Johnson, S. H., Matthew, D. B., Moulton, B. W., et al. (2002). Faith, confidence, and health care: Fostering trust in medicine through law, *1099*(2001).

Gilson, L. (2003). Trust and the development of health care as a social institution. *Social Science & Medicine, 56*(7), 1453–1468.

Goold, S. D. (2002). Trust, distrust and trustworthiness: Lessons from the field. *Journal of General Internal Medicine, 17*(1), 79.

Grande, D., Shea, J. A., & Armstrong, K. (2012). Pharmaceutical industry gifts to physicians: Patient beliefs and trust in physicians and the health care system. *Journal of General Internal Medicine, 27*(3), 274–279.

Haas, J. S., Phillips, K. A., Baker, L. C., Sonneborn, D., & McCulloch, C. E. (2003). Is the prevalence of gatekeeping in a community associated with individual trust in medical care? *Medical Care,* 660–668.

Halbert, C. H., Armstrong, K., Gandy, O. H., & Shaker, L. (2006). Racial differences in trust in health care providers. *Archives of Internal Medicine, 166*(8), 896–901.

Hall, M. A., Dugan, E., Balkrishnan, R., & Bradley, D. (2002). How disclosing HMO physician incentives affects trust. *Health Affairs, 21*(2), 197–206.

Hall, M. A., Dugan, E., Zheng, B., & Mishra, A. K. (2001). Trust in physicians and medical institutions: What is it, can it be measured, and does it matter? *Milbank Quarterly, 79*(4), 613–639.

Hall, M. A., Zheng, B., Dugan, E., Camacho, F., Kidd, K. E., Mishra, A., et al. (2002). Measuring patients' trust in their primary care providers. *Medical Care Research and Review, 59*(3), 293–318. https://doi.org/10.1177/1077558702059003004.

Hillen, M. A., De Haes, H. C. J. M., & Smets, E. M. A. (2011). Cancer patients' trust in their physician—A review. *Psycho-Oncology, 241*, 227–241.

Illingworth, P. (2002). Trust: The scarcest of medical resources. *The Journal of Medicine and Philosophy, 27*(1), 31–46.

Jindal, S. (1998). Privatisation of health care: New ethical dilemmas. *Issues Med Ethics,* 85–86.

Kao, A. C., Green, D. C., Davis, N. A., Koplan, J. P., & Cleary, P. D. (n.d.). Patients' trust in their physicians: Effects of choice, continuity, and payment method.

Kayaniyil, S., Gravely-Witte, S., Stewart, D. E., Higginson, L., Suskin, N., Alter, D., et al. (2009). Degree and correlates of patient trust in their cardiologist. *Journal of Evaluation in Clinical Practice, 15*(4), 634–640.

Keating, N. L., Green, D. C., Kao, A. C., Gazmararian, J. A., Wu, V. Y., & Cleary, P. D. (1997). How are patients' specific ambulatory care experiences related to trust, satisfaction, and considering changing physicians? *Journal of General Internal Medicine, 17*(1), 29–39.

Larson, H. J., & Heymann, D. L. (2010). Public health response to influenza A (H1N1) as an opportunity to build public trust. *JAMA, 303*(3), 271–272.

LaVeist, T. A., Nickerson, K. J., & Bowie, J. V. (2000). Attitudes about racism, medical mistrust, and satisfaction with care among African American and white cardiac patients. *Medical Care Research and Review, 57*(1_suppl), 146–161.

Lee, Y.-Y., & Lin, J. L. (2009). Trust but verify: The interactive effects of trust and autonomy preferences on health outcomes. *Health Care Analysis, 17*(3), 244–260.

Lee, Y., & Lin, J. L. (2009). The effects of trust in physician on self-efficacy, adherence and diabetes outcomes. *Social Science & Medicine, 68*(6), 1060–1068. https://doi.org/10.1016/j.socscimed.2008.12.033.

Lewicki, R. J., Mcallister, D. J., & Bies, R. J. (2008). Trust and distrust: New relationships and realities. *Academy of Management Review, 23*(3), 438–458.

Mainous, A. G., Baker, R., Love, M. M., Gray, D. P., & Gill, J. M. (2001). Continuity of care and trust in one's physician: Evidence from primary care in the United States and the United Kingdom. *Family Medicine, 33*(1), 22–27.

Makinen, M., Waters, H., Rauch, M., Almagambetova, N., Bitrán, R., Gilson, L., et al. (2000). Inequalities in health care use and expenditures: Empirical data from eight developing countries and countries in transition. *Bulletin of the World Health Organization, 78,* 55–65.

Mckinstry, B., Ashcroft, R., Car, J., Freeman, G., & Sheikh, A. (2008). Interventions for improving patients' trust in doctors and groups of doctors. *Cochrane Database of Systematic Reviews,* (3).

Mechanic, D. (1996a). Changing medical organization and the erosion of trust. *The Milbank Quarterly,* 171–189.

Mechanic, D. (1996b). The logic and limits of trust. *Contemporary Sociology, 25*(4), 455.

Mechanic, D. (1998). The functions and limitations of trust in the provision of medical care. *Journal of Health Politics, Policy and Law, 23*(4), 661–686.

Mechanic, D. (2001). The managed care backlash: Perceptions and rhetoric in health care policy and the potential for health care reform. *The Milbank Quarterly, 79*(1), 35–54.

Mechanic, D., & Meyer, S. (2000). Concepts of trust among patients with serious illness. *Social Science & Medicine, 51,* 657–668.

Misztal, B. (2013). *Trust in modern societies: The search for the bases of social order*. Wiley.

Mohseni, M., & Lindstrom, M. (2007). Social capital, trust in the health-care system and self-rated health: The role of access to health care in a population-based study. *Social Science & Medicine, 64*, 1373–1383. https://doi.org/10.1016/j.socscimed.2006.11.023.

Mollborn, S., Stepanikova, I., & Cook, K. S. (2005). Delayed care and unmet needs among health care system users: When does fiduciary trust in a physician matter? *Health Services Research, 40*(6p1), 1898–1917.

O'Neill, O. (2004). Informed consent and public health. *Philosophical Transactions of the Royal Society of London. Series B: Biological Sciences, 359*, 1133–1136. https://doi.org/10.1098/rstb.2004.1486.

Ohya, Y., Williams, H., Steptoe, A., Saito, H., Iikura, Y., Anderson, R., et al. (2001). Psychosocial factors and adherence to treatment advice in childhood atopic dermatitis. *Journal of Investigative Dermatology, 117*(4), 852–857.

Ozawa, S., & Walker, D. G. (2011). Comparison of trust in public vs private health care providers in rural Cambodia. *Health Policy and Planning, 26*, 20–29. https://doi.org/10.1093/heapol/czr045.

Pearson, S. D., & Raeke, L. H. (2000). Patients' trust in physicians: Many theories, few measures, and little data. *Journal of General Internal Medicine, 02215*.

Piette, J. D., Heisler, M., Krein, S., & Kerr, E. A. (2005). The role of patient-physician trust in moderating medication nonadherence due to cost pressures. *Archives of Internal Medicine, 165*(15), 1749–1755.

Plutchik, R. (2003). *Emotions and life: Perspectives from psychology, biology, and evolution*. American Psychological Association.

Rehman, S. U., Nietert, P. J., Cope, D. W., & Kilpatrick, A. O. (2005). What to wear today? Effect of doctor's attire on the trust and confidence of patients. *The American Journal of Medicine, 118*, 1279–1286. https://doi.org/10.1016/j.amjmed.2005.04.026.

Riewpaiboon, W., Chuengsatiansup, K., Gilson, L., & Tangcharoensathien, V. (2005). Private obstetric practice in a public hospital: Mythical trust in obstetric care. *Social Science & Medicine, 61*, 1408–1417. https://doi.org/10.1016/j.socscimed.2004.11.075.

Russell, S. (2005). Treatment-seeking behavior in urban Sri Lanka: Trusting the state, trusting private providers. *Social Science & Medicine, 61*, 1396–1407. https://doi.org/10.1016/j.socscimed.2004.11.077.

Safran, D. G., Taira, D. A., Rogers, W. H., Kosinski, M., Ware, J. E., & Tarlov, A. R. (1998). Linking primary care performance to outcomes of care. *Journal of Family Practice, 47,* 213–220.

Schneider, C., & Schneider, D. (1998). *The practice of autonomy: Patients, doctors, and medical decisions*. Oxford University Press on Demand.

Shore, D. A. (2003). Communicating in times of uncertainty: The need for trust. *Journal of Health Communication, 8*, 13–14. https://doi.org/10.1080/10810730390224839.

Straus, S. E., & McAlister, F. A. (2000). Evidence-based medicine: A commentary on common criticisms. *CMAJ, 163*(7), 837–841.

Thom, D. H. (2000). Training physicians to increase patient trust. *Journal of Evaluation in Clinical Practice, 6*(3), 245–253.

Thom, D. H. (2001). Physician behaviors that predict patient trust. *Journal of Family Practice, 50*(4), 323.

Thom, D. H., Hall, M. A., & Pawlson, L. G. (2004). Measuring patients' trust in physicians when assessing quality of care. *Health Affairs, 23*, 124–132. https://doi.org/10.1377/hlthaff.23.4.124.

Thom, D. H., Kravitz, R. L., Bell, R. A., Krupat, E., & Azari, R. (2002). Patient trust in the physician: Relationship to patient requests. *Family Practice, 19*(5), 476–483.

Thom, D. H., Ribisl, K. M., Stewart, A. L., Luke, D. A., & The Stanford Trust Study Physicians. (1999). Further validation and reliability testing of the Trust in Physician Scale. *Medical Care,* 510–517.

Thompson, M., Gee, S., Larson, P., Kotz, K., & Northrop, L. (2001). Health and loyalty promotion visits for new enrollees: Results of a randomized controlled trial. *Patient Education and Counseling, 42*(1), 53–65.

Trachtenberg, F., Dugan, E., & Hall, M. A. (2005). How patients' trust relates to their involvement in medical care. *Journal of Family Practice, 54*(4).

Van der Schee, E., Braun, B., Calnan, M., Schnee, M., & Groenewegen, P. P. (2007). Public trust in health care: A comparison of Germany, The Netherlands, and England and Wales. *Health Policy, 81*, 56–67. https://doi.org/10.1016/j.healthpol.2006.04.004.

Zaner, R. M. (1991). The phenomenon of trust and the patient-physician relationship. In *Ethics, trust, and the professions: Philosophical and cultural aspects* (pp. 45–65).

Chapter 2
Community-Based Exploratory Research on Trust in Physicians

Abstract In this chapter, I describe the various research methods I adopted to explore the idea of trust in physicians in the study setting. The study was conducted in Tamil Nadu, a state in southern part of India with good health indicators. The initial exploration was carried out using a mixed qualitative and quantitative method. The qualitative in-depth interviews were designed and conducted to understand the dimensions and determinants of trust in physicians in the developing healthcare settings. This was followed by a quantitative survey among systematically sampled populations from various parts of the state. After obtaining a conceptual understanding of trust in physicians using these two exploratory methods, psychometric methods were used to develop and validate a scale to measure trust in physicians in this setting. This was followed by a smaller survey to understand the perspectives of healthcare providers about trust in physicians.

Keywords Mixed methods research · Qualitative study · Quantitative study · Trust in physicians

This book describes the key findings of a community-based exploratory research on trust in physicians in a developing healthcare setting in India. The research was conducted between 2012 and 2014, in select districts of the south Indian state of Tamil Nadu. It was part of my doctoral dissertation. In this chapter, I describe the methods applied to conduct this research. The key objective of this research was to understand the concept of trust in physicians in developing healthcare settings. The primary research questions were:

- What are the dimensions of trust in physicians in such developing healthcare settings?
- What factors influence trust in physicians in these settings?

As these research questions were being answered during the initial stages of the research, a potential opportunity emerged to develop a scale to measure trust in physicians in the developing healthcare settings. Therefore, this study led to a psychometric exercise, in which a simple scale was developed to quantify trust in physicians in the developing healthcare settings and this scale was validated. Further, it was also found that there are substantial differences in the way trust in physicians

V. Gopichandran, *Dynamics of Trust in Doctor-Patient Relationship in India*,
SpringerBriefs in Ethics, https://doi.org/10.1007/978-981-15-0346-7_2

was perceived by patients and physicians themselves. Therefore, a quantitative survey was undertaken among a variety of healthcare providers to clearly understand their perceptions of trust in the physician–patient relationship.

This study adopted a mixed method sequential qualitative followed by quantitative research design. In the first phase, a qualitative exploration was performed to understand the dimensions of trust in physicians in these developing healthcare settings. Qualitative research helps answer the questions of 'why?', 'how?'. Therefore, qualitative methods are best suited to understand what people in the community mean by 'trust in physicians'. Who according to them is a trustworthy physician? How does trust build in the physician? How does trust erode? However, the qualitative research provides information that is true to the people who participated in the study. It is a form of co-creation of knowledge between the participants and researchers. The findings may not be generalizable to all people in the community. Anyways, generalizability is not the key objective in qualitative research. The important role of qualitative research is to understand the universe of truths related to trust in physicians. From a state in which we do not know how 'trust in physicians' functions in a typical developing healthcare setting, the qualitative research takes us to a state of emerging understanding about the concept of trust.

In order to systematically validate the identified concepts, a large-scale community-based quantitative survey was followed up immediately after the qualitative study. Quantitative surveys use structured questionnaires. Using these structured questionnaires, people's responses are documented. The objective of quantitative research is to find out how many people in the community responded in a certain way to each question asked in the questionnaire. This gives an idea of the common dimensions of trust and the less common ones. The objective of the quantitative survey was to identify the common truth about 'trust in physicians' in a specific population. The qualitative study informed this survey in that it provided the questions to be asked in the quantitative study. Based on the data obtained from this quantitative survey, several statistical methods were applied and the construct of trust in physicians was described in detail. These statistical procedures and methods are described in subsequent chapters.

At this point of time, we had a set of ideas which were identified by most people to be related to trust in physicians. This presented an opportunity to use these ideas in a manner such that it can measure the level of trust in physicians. This scale development process was undertaken, and the psychometric properties of trust in physician scale were assessed. The newly developed trust in physicians scale for the developing healthcare settings was validated using classical test and item response theory approaches both of which are described in detail in a later chapter. All these studies had looked at trust in physicians from the community perspective. Finally, a quantitative survey was also conducted among healthcare providers to triangulate the findings and understand the perspective of the physicians.

The following pages describe the methods adopted in detail, in order to place the findings that are described in the subsequent chapters in perspective (Table 2.1).

Table 2.1 Study design

Phase of study	Study objective	Specific objective	Study design	Process
Phase 1	To understand the concept of trust in physicians in developing healthcare settings	To explore the dimensions and determinants of trust in physicians	Qualitative study	In-depth interviews among 35 respondents
		To quantify and validate the identified dimensions and determinants of trust in physicians	Quantitative sample survey	Questionnaire administration on a sample of 616 adults
Phase 2	To develop and validate a scale to measure trust in physicians	To develop a new trust in physician scale using classical and item response theories	Quantitative sample survey	Classical test analysis and item response analysis
		To validate the new trust in physician scale	Quantitative sample survey	Administration of trust in physician scales and identifying its ability to discriminate people who have trust and those who do not
Phase 3	To identify factors perceived by health providers to influence trust in physicians		Quantitative sample survey	Questionnaire-based survey among 200 health care providers

Study Setting—Tamil Nadu, India

The study was conducted in Tamil Nadu, one of the coastal south Indian states forming the southernmost part of the country. India has a large public health system which runs through decentralized state budget allocation with contributions from the central government. With the advent of the National Rural Health Mission, a flagship health system strengthening program of the government of India in 2005, the public health system received a fillip in terms of decentralization, better platforms for community engagement with health care, better accountability mechanisms and greater fund allocation (Ramani & Mavalankar, 2006).

Alongside this strong public health system, there is also a powerful private sector in health care. Private health providers, who deliver healthcare services for a fee, are the highest contributors to health services in India. There is a growing corporate health industry in the metropolitan cities which provide international quality health services not only to the people in the country but also serve as hubs for health tourism. In addition to the private and public health systems, there is a big network of unorganized, unqualified medical practitioners providing all levels of health care (Peters, Rao, & Fryatt, 2003).

Tamil Nadu has a strong government-sponsored health insurance scheme in operation since 2009. Initially, it provided treatment in government and accredited private hospitals for secondary and tertiary care hospitalization and procedures up to an amount of Rs. 1 lakh per family per year. Now, this scheme has been integrated with the Prime Minister's National Health Protection Scheme (PMJAY) and it covers all people living below poverty line for cashless treatment of up to Rs. 5 lakhs per family per year. This scheme is a progressive step towards universal health coverage in the state (Lahariya, 2018).

In India, the overall health expenditure is about 4% of the Gross Domestic Product (GDP), and the government budget allocation for health care is less than 1% of the GDP. The remaining health expenditure is largely out of pocket. This leads to significant impoverishment, and catastrophic health expenditure is one of the commonest reasons for indebtedness in the country (Balabanova et al., 2013).

The public health system is plagued by system inefficiencies, irregularities, corruption and irrational health practices. The private providers largely remain unregulated and there is high prevalence of irrational care and commoditization of health care. Tamil Nadu is one of high performing states in India with respect to health indicators. It has one of the well-functioning models of health care in the country, but still several pockets especially poor rural areas and migrant urban populations remain largely underserved. The inefficient public health system, burgeoning private health care, rising cost of health care, irrational and unregulated practices make the study setting very different and unique from the other countries from where studies of trust in health care have previously emerged. Henceforth, we will refer to these health settings as developing health systems.

Need for Understanding Trust in Physicians in Developing Healthcare Settings

The developing healthcare settings have unique characteristics as described above. As seen earlier trust is a sense of forward-looking anticipation that the physician will do whatever is the right thing for the patient. We saw earlier that such anticipation largely rests on the trustworthiness of the physician. The trustworthiness of the doctor is likely to be strongly influenced by the medical practice environment. Therefore, the dynamics of trust in the physician–patient relationship is also likely to be highly contextual. Given the contextual nature of trust, its dimensions and determinants are likely to be different in developing healthcare contexts. Take for example, preserving patient autonomy as a dimension of trust in physicians. In a typical individualistic Western culture as in the USA, there is a lot of emphasis placed on the autonomy of the individual. In comparison, several studies done in Eastern cultures like in India and China have shown that autonomy is not conceptualized at the individual level, but at the level of families and communities. It can be often seen that treatment decisions are made collectively by the family and the community. Therefore, preserving the autonomy of the patient may not work out as a dimension of trust in the physician. In well educated and empowered communities, objective assessment of competence of the physician is often done by the patients by looking up the profile of the physicians in the website of the hospital where they work and making comparative judgments of treatment success rates of the physicians. On the other hand in developing healthcare settings like India, the strongest assessment of competence of a physician comes from 'word of mouth' from patients who have taken treatment from the physician and become better. Thus, 'trust in physicians' is likely to be very different in the context of developing healthcare settings.

Not only cultural differences mentioned above, but also social and political differences make trust in physicians different in these developing healthcare settings. In a setting where there is no universal access to health care, limited public spending on health and no guarantee for the quality of care received in the health system, it is likely that vulnerability will be more. Therefore, those who can afford the cost, may be able to pay to purchase 'trustworthiness' and those who cannot afford the cost may have to compromise for the care which they do not trust fully. This introduces a unique dynamic of trust in the physician–patient relationship. This kind of an unregulated and disturbed healthcare setting makes the community vulnerable to exploitation in the form of commercialization and commodification of health care. This can greatly influence the trust dynamics in the physician–patient relationship. Trust in physicians is therefore a very important concept to be studied in these developing health-care settings.

Qualitative Study to Identify Dimensions and Determinants of Trust in Physicians

The qualitative study was done to assess the various dimensions and determinants of trust in physicians as perceived by the community. Dimensions of trust are those components which formatively or reflectively form the construct of trust, whereas determinants are factors which influence whether a patient has high or low trust. This study aimed at assessing the differences between the dimensions and determinants of trust in physicians which has been previously described in developed health-care settings compared to the status in developing health systems like the study setting.

In-Depth Interviews

In-depth interview is a technique of qualitative data collection in which the interviewer asks open-ended questions and attempts to obtain information from the participant about their experiences, their opinions and perceptions. In-depth interviews help in understanding the depth of people's opinions and personal experiences. They help participants open and speak about personal issues, which they may not share in public forums (Methods, 2017).

Thirty-five in-depth interviews were conducted as part of this study, fifteen among migrant construction workers in and around Chennai, a metropolitan city in Tamil Nadu and sixteen among residents of a rural area in Dharmapuri district of Tamil Nadu and four interviews among primary care doctors catering to the health needs of these participants. Migrants and marginal subsistence rural farmers represent the extremes of the marginalized communities. Therefore, themes that emerge from their interviews are likely to shed more light on the aspects of trust in physicians in settings with severe resource deprivation.

The migrant construction labourers hail from various parts of India, largely from the states of Orissa, Uttar Pradesh and Jharkhand. They are marginalized because of language barriers, migrant status and poor living conditions provided in the construction sites. Health access to these migrant construction labourers is very poor. Dharmapuri is one of the districts of Tamil Nadu with poor health indicators. In the surveyed villages, the people are agricultural labourers and marginal subsistence farmers. In many households, the men and women of economically productive age group migrate outside the village to urban areas for work. Since Dharmapuri is a border district in the border between Tamil Nadu and its neighbouring state Karnataka, public services are compromised. Though the state of Tamil Nadu has some of the best health indicators in the country, Dharmapuri is among the poor performing districts of Tamil Nadu.

In-depth interviews were selected as the methodology for collecting data as the procedure is the most appropriate to gain insights into individual life experiences, trust and meaning ascribed to trust. I personally conducted all the interviews. Before going to perform my fieldwork, I underwent a rigorous training in the skills of conducting qualitative research interviews.

The construction labourers were approached in their place of residence, and the rural participants were interviewed during their visit to primary health centres in their respective villages for health care. The interviews in the rural area were conducted in Tamil language and the interviews with the migrant workers in Hindi.

I first started talking to the respondents about health and their perceptions of health. Then seamlessly, I led the interview towards illness, treatment seeking and choice of health facility. This was followed by discussion about trust in physicians. The various aspects of trust, what makes people trust the doctors and what makes people lose trust were explored during the interviews. The interviews last between 30 and 45 min each, with some interviews extending up to 2 h. The interviews were not recorded to prevent the respondents from becoming self-conscious, which often happens among marginalized communities who have an inherent mistrust for research. I took extensive notes of the interviews. On the same day, I came back to my place of stay and transcribed my notes along with all my observations of the interview as memos.

During my fieldwork in the construction sites, I also got the opportunity to closely observe the lives of the construction labourers, their housing sheds made of tin metallic sheets, their common unclean toilets, their cooking and eating practices and health-care seeking practices. All these observations were also noted down as memos during my qualitative research. Similarly, I stayed for an extended period in various villages in Dharmapuri during the interviews that I conducted there. Here again, I had the opportunity to closely observe the dynamics of health-care seeking and treatment seeking behaviours of patients in the primary health centres. In the rural community, my place of interview was the primary health centre in the villages and so most of my observations were based here.

Qualitative data analysis involved thorough reading of all the in-depth interview transcripts along with the observations and memos. Reading and immersing myself in the interviews was the first stage in the analysis. QSR NVivo software package version 7 was used for coding and analysis of the interviews. I read the notes several times and chose three information-dense interviews. I read the transcripts line by line and performed open coding, the process of assigning meanings to each of the phrases, sentences or paragraphs. The codes were verified and validated by my mentor and guide after discussions. Following this, I prepared a coding manual with detailed description of the codes. The remaining interviews were coded using this manual.

Due to the lack of availability of researchers trained in qualitative data analysis in the team, a third coder could not be engaged. In order to ensure an unbiased third set of coding and to have a fresh re-look at the data set for new perspectives, I packed up the data temporarily and immersed myself in other aspects of my work. I re-opened the data after a gap of 1 month and redid the coding of the interviews independent of the previous coding. The differences between the initial coding and recoding were identified and discussed with the second researcher till a consensus was arrived at. The codes were then grouped together into meaningful themes. The identification of themes was largely grounded in the data. But the influences of previous themes present in the literature on the identification of these themes cannot be ruled out. The main themes and their interrelationships were assessed based on the interviews. After identifying the themes, the conceptual framework was developed and discussed between my guide and me.

I am a medical doctor by training and so in some of the early interviews could have brought the bias into the interpretation of the results. For example, in some of the early interviews when negative behaviours of doctors and health personnel were pointed out, I found myself becoming defensive and this could have expressed in my body language and influenced the way the interview proceeded. I maintained a reflexivity journal to note down these changes that I observed in my interviews. In some interviews in Dharmapuri in the primary health centres, I found that the participant would suddenly become self-conscious while talking to me. Then, I realized that this was because they noticed a staff from the primary health centre walking by, and they did not want them overhearing the conversation. These were reflected upon and appropriately addressed as memos during the analysis.

The quantitative survey was conducted to confirm the dimensions and determinants of trust in physicians obtained in the qualitative study. The main aim of this study was to validate the various dimensions and determinants obtained in the qualitative study and to segment the community based on the factors that influence trust in physicians.

A sample of 625 adults living in four districts of Tamil Nadu was selected by a multistage sampling methodology. The sample size was calculated according to the heuristics for sampling in multivariate modelling that there should be at least 20 observations per variable of analysis (Lwanga & Lemeshow, 1991). Four districts were first selected by simple random sampling method from the 32 districts of Tamil Nadu. Two of these districts were predominantly urban, and two were predominantly rural. Three urban wards and eight rural blocks were selected from each district by probability proportion to size method. From each selected block/ward, 50 individuals, who reported having a regular primary care physician, were randomly interviewed for the study. Table 2.2 shows the sampling strategy followed in the study.

Based on the dimensions of trust that were identified in the qualitative study described previously, items were generated to address each of the dimensions. Factors that influence trust in physicians were also identified from the qualitative study, and questionnaire items developed from these factors were included in the questionnaire. The questionnaire items were circulated to ten people, three physicians, two public health professionals and five lay persons to get their opinion on the face validity of the questionnaire. They were asked to rate each question on a scale of 1–5 on the extent to which they represent the dimensions and determinants of trust in physicians

Table 2.2 Samples selected for the quantitative survey on trust in physicians

S. No	District	Block	Type of area	Villages	Sample
1	Tiruvallur	Sozhavaram	Rural	Aangadu	50
			Urban	Sozhavaram	50
			Rural	Panchetti	50
			Rural	Kummanur	50
2	Kancheepuram	Tirukazhukundram	Rural	Karumarapakkam	50
			Rural	Nerumbur	50
			Rural	Irumbulicheri	50
			Rural	Naduvakkarai	50
3	Dharmapuri	Palacode	Urban	Palacode	50
			Rural	Mannadi	50
4	Chennai	–	Urban	–	125
				Total	625

with score 1 being least representative and 5 being most representative. The items which scored least were removed from the final scale.

The questionnaire was developed in English, and then, I translated it to the local language Tamil. A neutral third person back-translated the Tamil version of the questionnaire to English. This back-translated version was checked for validity of the translation. Appropriate changes to the translation were made to ensure linguistic validity of the questionnaire.

The selected participants were requested to answer the questions based on the primary care physician that they trusted the most for their minor ailments. They answered the questions on attributes of trust with respect to this physician. Further, they also answered questions on the various factors influencing trust with respect to this physician. For example, I would state 'I get the confidence that all my illness will get alright when I go to the doctor' (domain—perceived competence), 'The doctor gives appropriate medications for my diseases' (domain—perceived competence) reflecting attributes of trust. Further, I would state 'the doctor explains to me clearly about the illness and its treatment' (domain—physician behaviour), 'I do not have any inhibitions with him/her' (domain—comfort). The respondents rated the statements on a five-point Likert scale between 'Strongly agree' to 'Strongly disagree'.

Hospital-based sampling was avoided. I went door to door to administer the questionnaires to the participants. I had the support of two other trained investigators. Though I am a physician by profession, this was not made explicit to the respondents during the interviews in order to avoid reporting bias. By using a standardized questionnaire and standardization of the questionnaire administration process, the chance of interviewer bias was controlled.

The data was managed using SPSS Statistical Software version 17.0.1. Simple descriptive statistics such as frequencies of discrete variables, means and standard deviations of continuous variables were computed using the software package. Factor analysis was used to identify the components of factors influencing trust in physicians. Hierarchical cluster analysis with the factor scores was used to determine the number of clusters that the sample could potentially be segmented into. Finally, the factor scores were used to segment the respondents into groups by k-means cluster analysis. The characteristics of each segment were identified using descriptive analysis.

In order to validate the conceptual framework obtained in the qualitative study, a hypothetical model was fit based on this framework. This included trust in physicians as the key latent variable. Twelve relevant items indicating 'perceived competence', 'treatment assurance', 'confidence', 'respect' and 'loyalty' domains of trust were included as reflective variables contributing to the latent variable trust. Factors influencing trust in physicians were further added to the structural model as independent latent variables contributing to the dependent latent variable 'trust in physicians'. These factors were 'shared identity', 'behaviour of the physician', 'personal involvement of physician' and 'comfort with the physician'. Other variables such as age, sex, education, occupation and area of residence were also included in the model as endogenous variables.

Before conducting the measurements, important assumptions of structural equation modelling were considered. Conditions for causality were not met in this data as it was a cross-sectional assessment and temporality was not demonstrated. Only the factors which were identified in the qualitative exploration were included in the model. Other confounding variables or unknown factors could be involved in the model, which were not identified or included. The direction of the proposed causal relationship is also not exclusive. For example, it is possible that comfort levels would increase trust and at the same time increased trust could also contribute to greater comfort levels. Despite these limitations in the conceptual assumptions of the structural equation modelling, the model was fit to explore hypothetical associations. Nevertheless, these limitations have to be borne in mind while interpreting the results. The data was in an ordinal Likert-type scale. It was assumed to be continuous for analysis purposes. The variables were checked for normality and linearity. Based on the acceptable skewness and kurtosis values suggested by West et al. for fitting structural equation models with non-normal data, the model was assumed to be normal. The model fit statistics were computed with this assumption of normality. The standardized root mean square residual (SRMR), root mean square error approximation (RMSEA) and comparative fit index (CFI) were used to assess model fit. SRMR of less than 0.09, RMSEA of less than 0.06 and CFI of more than 0.90 were considered as adequate model fit criteria (MacCallum & Austin, 2000). To get more robust regression estimates, bootstrapping was done and the confidence intervals of the estimates were calculated using the bias-corrected percentile method. The Bollen–Stine bootstrap model fit chi-square was also calculated to further test for model fit. After the construction of the structural equation model, two measures were adopted to improve the model fit. Firstly, variables which did not contribute significantly to the factor, both conceptually and statistically, were sequentially removed. Secondly, meaningful error covariances were introduced between variables in the same factor based on modification indices reported by the program.

Psychometric Scale Development Study

Following the understanding of the dimensions and determinants of trust in physicians using qualitative and quantitative assessments, a series of studies were conducted to develop and validate a scale to measure trust in physicians in the local context.

The data collected for the quantitative community survey was used for this purpose. The method of data collection has been described previously. The data was managed using SPSS Statistics version 17.0.1, IBM SPSS Amos version 20 and item response theory for patient-reported outcomes (IRTPRO) version 2.1. All the 31 items in the questionnaire relating to the dimensions of trust in physician were assessed in this analysis. Classical test analysis of the items was done by performing internal consistency test using Cronbach's alpha, item-to-total correlation and interitem correlations. Confirmatory factor analysis (CFA) was done to assess dimensionality of selected well performing twenty-two items from the total of thirty-one. Standardized root mean squared residuals (SRMR) of less than 0.08, comparative fit index (CFI) greater than 0.95, Tucker–Lewis index (TLI) greater than 0.95 and root mean square error approximation (RMSEA) less than 0.06 were used for assessing fitness of the CFA model. Local dependence was checked by performing the bifactor LD-X2 values considering a value of greater than 10 to indicate local dependence. After confirming unidimensionality and local independence, Samejima's graded response model was fit for the selected 22 questions (Drasgow & Parsons, 1983). Two models were fit, a one-parameter logistic model (1PL) and a two-parameter logistic model (2PL). The Akaike information criterion (AIC) and Bayesian information criterion (BIC) and −2log likelihood were computed for the two models. Based on assessment of the classical test properties and the item parameters, a final scale was identified selecting the optimally performing items.

For further predictive validation of the scale, a study was done to assess the ability of the new trust in physician scale to classify individuals as trusting and non-trusting of their physicians. An urban residential area in Chennai, Tamil Nadu, was randomly selected for the study. From a random start in this area, every fifth household was chosen till a total sample of 234 households was obtained. In the first sampled household, one adult respondent was randomly selected and asked to think of a doctor whom they trusted most and respond to the trust in physician questionnaire based on this doctor. In next sampled household, the respondent was asked to think of a doctor whom they did not trust and respond to the trust in physician questionnaire based on this doctor. If a household did not have any adult respondents available, was locked or refused to participate in the study, the household was not replaced. Apart from the trust in physician questionnaire, basic demographic information was also collected. Independent t-test was done to assess the difference in the mean scores for each item between the trusting and distrusting groups. A discriminant function model was fit to assess the effectiveness of the classification of the trusting and distrusting groups based on the questionnaire.

Quantitative Survey Among Healthcare Providers About Trust in Physicians

In order to understand the perceptions of healthcare providers on factors influencing trust in physician, a study was done in Kerala, the neighbouring state of Tamil Nadu, where the perspectives of healthcare providers were studied. The study was done in seven healthcare facilities, hospitals, clinics and community health centres in Ernakulum and Trivandrum districts of Kerala. The samples for the study included doctors belonging to different specialties, dentists, nurses, pharmacists and paramedical staff.

The instrument used for data collection was a questionnaire which was prepared specifically for the purpose of the study. For the ease of data collection, both English and Malayalam (the local language) versions of the questionnaire were used. The instrument aimed at assessing the perception of healthcare providers regarding their own trustworthiness. The respondents were asked to provide their basic information including age, gender, designation, type of hospitals where they work, degrees they hold and their years of experience. It was followed by a research question—'When do you think that the patient will trust you more'? The respondents were given a questionnaire containing 62 statements and were asked to rate their perceptions in a 3-point Likert scale of 1–3 where 1 stands for 'agree', 2 stands for 'neutral' and 3 stands for 'disagree'. The statements were representing various domains of the theme—trustworthiness.

All healthcare providers who were currently working in any preventive or curative settings were eligible to participate in the study. The sample size was fixed at 200 for sake of convenience. Written informed consent was obtained from the participants before the study. The respondents were provided adequate privacy for answering the questions, and they were assured that their responses would be kept confidential. The researcher, who was not part of the hospital treatment team, administered the questionnaire in Malayalam language or English language and gave the respondents enough time to think and answer each question. After the respondent gave the answers, they were noted down by the researcher.

The data was managed using SPSS Statistics version 17.0.1. After the data cleaning, only 188 complete samples were selected for further analysis as the remaining 12 questionnaires had more than 20% missing responses.

In summary, this study represents the opinions and perspectives of roughly 1000 individuals living in various districts of Tamil Nadu and its neighbouring state Kerala in south India. The experiences, perspectives and opinions of these 1000 people can be said to be representative of a typical developing health-care setting. Therefore, the findings are fairly generalization to a context of health care as described previously as a developing healthcare setting. In the following chapters, the important findings of the study will be discussed in detail.

References

Balabanova, D., Mills, A., Conteh, L., Akkazieva, B., Banteyerga, H., Dash, U., et al. (2013). Good health at low cost 25 years on: Lessons for the future of health systems strengthening. *The Lancet, 381*(9883), 2118–2133.

Drasgow, F., & Parsons, C. K. (1983). Application of unidimensional item response theory models to multidimensional data. *Applied Psychological Measurement, 7*(2), 189–199. https://doi.org/10.1177/014662168300700207.

Lahariya, C. (2018). 'Ayushman Bharat' program and universal health coverage in India. *Indian Pediatrics, 55*(6), 495–506.

Lwanga, S. K., & Lemeshow, S. (1991). *Sample size determination in health studies: A practical manual*. World Health Organization.

MacCallum, R. C., & Austin, J. T. (2000). Applications of structural equation modeling in psychological research. *Annual review of psychology, 51*(1), 201–226.

Green, J., & Thorogood, N. (2004). *Qualitative Methods for Health Research*. London: Sage Publications.

Peters, D. H., Rao, K. S., & Fryatt, R. (2003). Lumping and splitting: The health policy agenda in India. *Health Policy and Planning, 18*(3), 249–260. https://doi.org/10.1093/heapol/czg031.

Ramani, K. V., & Mavalankar, D. (2006). Health system in India: Opportunities and challenges for improvements. *Journal of Health Organization and Management, 20*(6), 560–572.

Chapter 3
Meaning of Trust in Physicians in Developing Healthcare Settings

Abstract In the setting where this study was conducted, the key dimensions of trust in physicians were perceived competence of the physician, assurance of treatment, willingness to accept drawbacks in the physician–patient relationship, loyalty towards the physician and respect. Based on these identified dimensions, a questionnaire was developed, and a survey conducted. The results of the survey indicated the most common dimensions were perceived competence, assurance of treatment, loyalty and respect. The various results and their descriptions are presented in detail in this chapter.

Keywords Trust in physicians · Perceived competence · Assurance of treatment · Respect · Loyalty

The primary goal of this research was to understand the meanings attributed by the people to trust in physicians. In this chapter, I will attempt to describe the findings of the qualitative study as well as the quantitative exploration of the meaning of trust in physicians as mentioned by the participants. The meaning of a phenomenon like trust in physicians is best understood by its component dimensions. These dimensions may be formative or reflective. A formative concept of trust is one in which each dimension adds up to create the whole idea of trust. On the other hand, a reflective concept of trust is one in which the changes in trust will reflect as changes in each of the component dimensions. As a result of a detailed exploration using the qualitative methods, the various dimensions of trust were identified. Then these dimensions were analysed and classified as those that were formative or reflective.

What Are the Dimensions of Trust in Physicians in Developing Healthcare Settings?

When asked about what people meant by trust in physicians, they described various dimensions.

V. Gopichandran, *Dynamics of Trust in Doctor-Patient Relationship in India*, SpringerBriefs in Ethics, https://doi.org/10.1007/978-981-15-0346-7_3

Perceived Competence

The first and foremost dimension described by them was 'perceived competence'. The community had some mechanisms for assessing the competence of the physician. In the typical collectivist type of society in the study area, the community perception of competence of a physician was shared by the community. Positive treatment experiences were immediately shared in the community and the members of the community placed trust in the physician. Word of mouth was the most powerful medium of sharing community perceptions of competence of the physician. It was seen that when a physician in a primary health centre in the villages of Dharmapuri, was providing good-quality clinical care, the physician became a household name in the local area and trust levels increased. One of the definitions of trust explains that while confidence is a metric of competence, trust is a metric of integrity, fidelity, honesty and benevolence. It is important to understand how competence, integrity, fidelity and benevolence assessments are performed by the community in developing healthcare settings. This may give a clue to why in these settings a perception of competence is a dimension of trust.

The perception of competence in a low literacy, low health awareness, high vulnerability setting, is seldom based on assessment of technical competence in the form of educational qualification, place of training, level of skills or expertise. The competence is usually perceived based on two major aspects—positive health outcomes and avoidance of medical errors. In one of the field interviews in Dharmapuri, an elderly person mentioned that he had knee pain due to arthritis. He was visiting a primary health centre regularly for the treatment. He said that the physician in the primary health centre gave him medicines that relieved his pain, and that was what mattered the most. The elderly man seemed not to be bothered by the fact that the physician in the primary health centre was not an orthopaedic surgeon, who is the specialist to whom one would go for a bone or join problem.

Another important dimension of perception of competence of the physician was fulfilment of certain basic expectations that the patients had from their physicians. In developing healthcare settings, it is very common for patients to expect an injection during a clinic visit. In some cases, the people also expect intravenous infusion of fluids. Fulfilment of these expectations was a very strong dimension of trust. It was common to hear the narrative of 'that physician always gives a good injection or glucose drips, which immediately makes us feel better'—as a reason for trusting the physician. In this context, rationality of the treatment is not of primary concern. In some cases, people also referred to those physicians who performed a thorough clinical examination, using their instruments such as a blood pressure apparatus and a stethoscope as competent physicians. In a busy outpatient clinic with about 200 patients being seen by one physician within a period of 4–5 h, such a thorough physical examination becomes a luxury and many patients do not receive that from their physician. However, patients do expect their physicians to examine them thoroughly.

So, when the expectations of the patients in the form of either a thorough clinical examination or giving an injection or intravenous fluid infusion are met, the trust in the physician builds.

The community also perceived different levels of competence of the physician for different illnesses. For example, the physician in the primary care centre was seen as competent in issues of health of mother and child including delivery, childcare, breastfeeding, contraception and so on. One could attribute this to the high emphasis placed by the government health system on maternal and child health care in Tamil Nadu.

While, on one hand, there was a perception of competence based on experience, in situations where there was no experience to fall back on, the competence judgment was more objectively judged by attributes that were obvious. A physician practicing in an expensive, posh looking, private health facility was seen as more competent than one working in a simple sparsely decorated primary health centre. A physician charging more fees was seen as someone obviously 'more important' and more specialized than one who was charging less or not charging any fee at all. A physician using technology such as computers, laboratory tests, scans and imaging services was seen as more competent than one who was relying on clinical skills alone. But all these types of competence judgments were in situations where there were no individual or collective positive experiences to rely on. The idea of perceived competence that emerged from this study was that in developing healthcare settings, there is still a significant role for word out mouth as a shared perception of competence. Communities tend to place their trust in physicians who are perceived as competent by their shared judgment, which is based on positive clinical experiences, rather than objective judgments. When there is no clinical experience to fall back on, certain external factors such as ambience, fees, use of technology play a role in the perception of competence.

Assurance of Treatment

In settings of resource limitation and lack of universal access to health care, the greatest vulnerability that people experience is during times of illness. It is well known that healthcare expenditure is the commonest cause of impoverishment in many low- and middle-income countries. A typical poor family living in a developing healthcare setting barely has resources to meet their everyday life expenses. So when they are faced by a health crisis, it leads to serious indebtedness. It pushes families into extreme poverty, from which many families never recover. In this kind of a setting, a basic minimum assurance of treatment at the time of illness is a significant dimension of trust in physician. One of the narratives that came up strongly during the interviews was that when their children are sick, the physician is readily available and can provide basic treatment and therefore they trust the physician. The availability of the physician in itself is seen as an assurance of treatment and therefore as a matter of trustworthiness. This dimension is a direct consequence of the health human resource crisis in many places with developing healthcare systems. A large number of public health facilities in India suffer from vacant positions of medical officers. Therefore, there is no assurance of receiving basic primary care for large groups of people. In this context, assurance of treatment was articulated as a major dimension of trust in the physician.

In settings with poverty, ability to pay out of pocket for treatment is often a luxury as seen earlier. Therefore, assurance of treatment irrespective of the ability to pay came across as a major dimension of trust. One of the villagers in the interviews narrated an instance where he had to take a friend of his for an emergency treatment to the primary health centre. The friend had sustained a fall from a height and had injured his leg, which had a deep cut and was bleeding. He explained the stress during the situation and how the physician in the primary health centre sutured the wound without demanding any money. He said that this greatly increased his trust in the physician. He recalled that how during a previous instance he had gone to the private physician in a nearby town and was turned away without any treatment because he could not afford to pay for it.

In one of the interviews, trustworthiness was also ascribed to the physician being available at any time of the day, in case of emergencies. The dedication, altruistic attitude and commitment of the physician who was available at any time of the day was respected by people and seen as a dimension of trust.

Another person very clearly described treatment from a government primary health centre as ‘home food’ and treatment in a private health facility as ‘eating out’. This metaphor referred to a sense of assurance that is there in home food versus the expensive indulgence that is associated with eating out. The emotional narrative very clearly explained the importance of universal health coverage in the trust dynamics in the physician–patient relationship.

In summary, a basic assurance of treatment, irrespective of the ability to pay, irrespective of the time of the day, was a major dimension of trust in the physician.

Willingness to Accept Limitations in the Physician–Patient Interaction

An interesting point that could be noted in the interviews was that judgments of perceived competence and assurance of treatment outweighed the judgment of other aspects of the physician–patient relationship such as kindness, politeness, and other communication skills. In most of the interviews, it was evident that the people were willing to pardon their physician for their rudeness, their impatience with the patient, scolding them in a paternalistic manner for their mistakes, and making them wait for long hours in the clinic, as long as they perceived the physician to be competent and were assured of a basic level of treatment.

In a developing healthcare setting, behavioural characteristics of the physician such as communication skills, respect to the patient’s agency and personhood did not emerge as dimensions of trust in physicians. These are probably background variables and only factors influencing the trust. None of the participants in the interviews mentioned kindness, compassion, politeness, etc., as dimensions of trust. They did not say these were unnecessary. But these were not the dimensions of trust.

In the primary health centre, often the physician saw patients in a hurried manner, barely giving a few minutes for each patient because of the huge crowd of patients. Often the physician came across as a rude person, with no time for politeness or niceties. Despite this, patients trusted their physicians based on their perception of competence of the physician and the fact that a basic level of treatment was assured.

One of the patients said that he was willing to go a private physician even though he had to spend a significant portion of his monthly wage on his treatment there because the treatment outcome was good. There the perception of competence and availability of treatment even outweighed the cost factor.

To an extent, the willingness to accept the limitations of the physician–patient relationship was also a reflective dimension of trust. The greater the trust in the physicians, the greater the willingness to accept the limitations in the physician–patient relationship.

Loyalty

Loyalty is a strong, blind and unquestioning sense of affiliation to a person or a group. Many patients in the developing healthcare setting described loyalty as a dimension of trust. They perceived a strong sense of affiliation to the physician or the health facility where the physician was working. In many interviews, the participants expressed their loyalty to the physician in very clear terms. They explained that they never went to see another physician after having found this one. They described that irrespective of the nature of illness, they always went to their trusted physician first. They also explained that even if they were referred to a specialist centre, they always made sure to come back to their trusted physician to verify the treatment before taking the treatment.

One of the patients said that he had suffered a heart attack a few years ago and was admitted in a tertiary care hospital to undergo a coronary bypass surgery. He had insisted that the surgeon speaks to his primary care physician before the surgery and get his permission. It turned out the primary care physician was an unlicensed and unregistered practitioner in the village. The situation had ended up as a comedy of errors. However, the patient refused to mistrust his 'physician' despite this discovery because his trust was strongly based on his perceived competence of the 'physician' and the assurance of treatment that he had provided him for so many years. The greater the trust in the physician, the greater was the level of loyalty.

Respect

In a typical developing healthcare setting with high levels of illiteracy and low health awareness and poverty, education is perceived as a great power. The educated are highly respected in the society. Every poor man and woman aspire to educate their children, as it is viewed as the key to a better quality of life. The physician is seen as a highly learned person. The physician's education and professional status offers her a social status which commands respect. Even before there is any interaction with the physician, before any kind of perception of competence happens, there is an inherent respect in physicians. Subsequently, greater the level of trust in the physician, greater the respect.

From the above discussion, it is clear that the key dimensions of trust in physicians in a typical developing healthcare setting are perception of competence, assurance of treatment, willingness to accept limitations in the physician–patient relationship, loyalty and respect. There are some dimensions of trust in physicians that have been previously described from Western literature that are conspicuous by their absence in the discourse in this developing healthcare setting. Hall et al., described the dimension of fidelity (Hall, Dugan, Zheng, & Mishra, 2001). Under this dimension, they mentioned caring, avoiding conflicts of interest and expressing a genuine interest in the life of the patient. While patients in this study setting depended on a collective community experience and opinion about the outcomes of the treatment, they did not seem to make micro-level judgments about the intentions of the physician. In a setting of high level of vulnerability, such an assessment of the intentions of the physicians seems to be a luxury that they cannot afford. Hall et al., described competence as a key dimension of trust in physicians and they articulated it as communication competence. Listening to the patient, talking to them and explaining the details of the illness to them, politeness and kindness were articulated as the key dimensions of competence. However, this was the direct opposite of how trust is articulated in a developing healthcare setting. Competence assessment is based on positive treatment outcomes of the community. The other important trust dimension that was absent in this study was honesty. Once again, this is because honesty assessment is something that happens after a basic assurance of treatment is provided. Mechanic and colleagues described that people usually cannot arrange the dimension of trust in their order of importance (Mechanic, 1998). In this study, it was clearly arranged by the participants in order of importance. The most important dimensions were basic assurance of a treatment, perception of competence and these outweighed the importance of communication skills and dynamics of physician–patient relationship, which were viewed as luxuries.

Quantitative Survey Findings

Using these important dimensions of trust that were identified in the qualitative study, a quantitative survey was conducted using a structured questionnaire explained in a previous chapter.

Out of a total of 625 survey responses, 9 were incomplete and hence were removed from analysis. The characteristics of the remaining 616 respondents are shown in Table 3.1. The characteristics of the sample are compared with population characteristics of Tamil Nadu (Department of Statistics, 2014; Registrar General, 2011).

Table 3.1 Characteristics of the quantitative survey participants

Characteristic	Categories	Frequency (percentage)	Population characteristics of Tamil Nadu state
Age	18–25	158 (25.3%)	19%
	26–35	204 (32.6%)	16.7%
	36–45	104 (16.6%)	14%
	46–55	85 (13.6%)	10%
	56–65	45 (7.2%)	7%
	>65	29 (4.6%)	5%
Sex	Male	252 (40.9%)	50.2%
	Female	364 (59.1%)	49.8%
Place of residence	Urban	124 (20.1%)	48.45%
	Rural	501 (79.9%)	51.55%
Education	No formal education	89 (14.4%)	Literate—73.8%
	Schooling	270 (43.8%)	Illiterate—26.2%
	Graduation	135 (21.9%)	
	PG and professional	122 (19.8%)	
Occupation	Service sector	165 (26.8%)	22%
	Homemaker	159 (25.9%)	*
	Labourer	92 (15%)	26%
	Unemployed	72 (11.7%)	*
	Professional	42 (6.8%)	8%
	Agricultural land owner	38 (6.2%)	40%
	Business	30 (4.9%)	8%
	Skilled workers	17 (2.8%)	*
Any sickness in the past three months	Yes	323 (52.5%)	
	No	292 (47.5%)	

* Stands for "Data not available"

The frequencies of responses to the dimensions of trust in physicians are shown in Table 3.2. It is seen that perceived competence, treatment assurance, respect and loyalty are identified by most of the respondents as dimension of trust in physicians. Willingness to accept drawbacks is a less preferred dimension. The least preferred dimension of trust in physicians is confidentiality.

This finding validated the importance of perceived competence as highlighted in the qualitative study. Even among the competence perception, the responses indicate that competence perceived by self is given greater priority than shared competence assessments from the community. Apart from competence, assurance of treatment was the second-most commonly reported dimension of trust in physicians. In the developing health system setting, there is a huge uncertainty about access to good-quality health care in the public health system. In the private system, the problems of high cost of care and lack of standardized treatments create doubts. In this situation of uncertainty, assurance of treatment of acceptable quality, irrespective of ability to pay was seen as important dimension of trust. This dimension is unique to settings where there is no universal access to health care and out of pocket expenditure is very high.

Respect to the physician is another dimension which emerged as highly significant in the survey. Trust could also transform into blind faith in the presence of high levels of respect. Loyalty was also a key dimension of trust in the physicians. Respect and loyalty are emotional dimensions of trust. Both respect and loyalty are effects of trust in the physician. In that case, they are reflective dimensions of trust. Reflective dimension of trust means that when there is a change in the levels of trust, it is reflected as change in the levels of respect and loyalty.

Willingness to accept drawbacks in the physician, though it emerged as a dimension of trust in the qualitative study, did not play an important role in this quantitative exploration. This could be due to several important reasons. Firstly, the method of eliciting the response could have played a role in the discrepancy. When asked whether the patient will tolerate if the doctor is rude to them, the response is less likely to be affirmative. On the other hand, it is more likely to be self-reported by a patient that they are willing to accept rudeness if the doctor is competent. Secondly, it is possible that the respondents interpreted the 'willingness to accept drawbacks' dimension as weakness and hence gave socially desirable responses to the statements. Finally, it is also possible that the respondents actually felt that their doctors were never rude, never charged them too much money or never made any mistakes in their treatment and therefore the statements on willingness to accept the drawbacks were not relevant to them. This dimension of trust was not carried forward any further for analysis in the subsequent sections. The quantitative survey confirmed four important dimensions of trust in physicians, namely perceived competence, treatment assurance, respect and loyalty.

Table 3.2 Responses to the dimensions of trust in physicians

Questionnaire items	Strongly agree *n* (%)	Agree *n* (%)	Neither agree nor disagree *n* (%)	Disagree *n* (%)	Strongly disagree *n* (%)
The physician gives appropriate medicines for my disease	439 (71.3)	84 (13.6)	38 (6.2)	31 (5)	24 (3.9)
If I go to the physician, I will surely get good treatment for my illness	428 (69.5)	88 (14.3)	45 (7.3)	31 (5)	24 (3.9)
The physician relieves the illnesses quickly	407 (66.1)	118 (19.2)	38 (6.2)	25 (4.1)	28 (4.5)
I respect this physician a lot	406 (65.9)	105 (17)	55 (8.9)	19 (3.1)	31 (5)
The physician prescribes appropriate number of medicines based on the nature of the illness	405 (65.7)	85 (13.8)	73 (11.9)	21 (3.4)	32 (5.2)
There is a big crowd in the clinic of the physician	397 (64.4)	88 (14.3)	78 (12.7)	27 (4.4)	26 (4.2)
The physician does important blood tests and other tests to diagnose my disease	385 (62.5)	88 (14.3)	61 (9.9)	40 (6.5)	42 (6.8)

(continued)

Table 3.2 (continued)

Questionnaire items	Strongly agree *n* (%)	Agree *n* (%)	Neither agree nor disagree *n* (%)	Disagree *n* (%)	Strongly disagree *n* (%)
I think this physician is a very learned person	379 (61.5)	108 (17.5)	67 (10.9)	22 (3.6)	40 (6.5)
There are no side effects to the medicines prescribed by the physician	366 (59.4)	95 (15.4)	69 (11.2)	34 (5.5)	52 (8.4)
The main intention of the physician is to cure my illness and not anything else	364 (59.1)	104 (16.9)	70 (11.4)	35 (5.7)	43 (7)
Friends, relatives and neighbours recommend me to go to the physician	359 (58.3)	124 (20.1)	64 (10.4)	24 (3.9)	45 (7.3)
I get the confidence that all my illness will get alright when I go to the physician	352 (57.1)	135 (21.9)	72 (11.7)	24 (3.9)	33 (5.4)
Friends, relatives and neighbours speak well about the treatment provided by the physician	350 (56.8)	114 (18.5)	84 (13.6)	27 (4.4)	41 (6.7)
I will bring my family members only to this physician	348 (56.5)	90 (14.6)	86 (14)	29 (4.7)	63 (10.2)

(continued)

Table 3.2 (continued)

Questionnaire items	Strongly agree *n* (%)	Agree *n* (%)	Neither agree nor disagree *n* (%)	Disagree *n* (%)	Strongly disagree *n* (%)
The illness gets cured with just one visit, there is no need for repeat visits	346 (56.2)	121 (19.6)	76 (12.3)	29 (4.7)	44 (7.1)
I will recommend only this physician to all those who ask me	345 (56)	100 (16.2)	80 (13)	34 (5.5)	57 (9.3)
The physician gives me good treatment irrespective of whether I have money to pay	320 (51.9)	81 (13.1)	88 (14.3)	42 (6.8)	85 (13.8)
Whatever illness I have, I will go only to this physician	320 (51.9)	75 (12.2)	82 (13.3)	53 (8.6)	86 (14)
Whether it is personal issues or health-related issues I will ask this physician for advice	307 (49.8)	85 (13.8)	83 (13.5)	44 (7.1)	97 (15.7)
I admire this physician	307 (49.8)	101 (16.4)	91 (14.8)	30 (4.9)	87 (14.1)
I ask for second opinion or reconfirmation of the physician's treatment	301 (48.9)	70 (11.4)	85 (13.8)	63 (10.2)	97 (15.7)

(continued)

Table 3.2 (continued)

Questionnaire items	Strongly agree *n* (%)	Agree *n* (%)	Neither agree nor disagree *n* (%)	Disagree *n* (%)	Strongly disagree *n* (%)
I will not accept it if anybody points out mistakes of the physician, irrespective of whether it is true or false	299 (48.5)	85 (13.8)	129 (20.9)	37 (6)	66 (10.7)
For serious illnesses more expensive medicines are prescribed	296 (48.1)	112 (18.2)	105 (17)	39 (6.3)	64 (10.4)
I sometimes doubt if the physician has the ability to treat problems like mine	296 (48.1)	69 (11.2)	79 (12.8)	75 (12.2)	97 (15.7)
Irrespective of what time of the day it is, whenever I go, I can get good treatment with the physician	285 (46.3)	99 (16.1)	83 (13.5)	56 (9.1)	93 (15.1)
Even if I go to another physician, I will take the treatment only if this physician approves it	253 (41.1)	75 (12.2)	82 (13.3)	53 (8.6)	86 (14)

(continued)

Table 3.2 (continued)

Questionnaire items	Strongly agree n (%)	Agree n (%)	Neither agree nor disagree n (%)	Disagree n (%)	Strongly disagree n (%)
I will tolerate it even if the physician takes huge amounts of money from me	143 (23.2)	67 (10.9)	122 (19.8)	68 (11.2)	215 (34.9)
I will tolerate it even if the physician behaves rudely with me	139 (22.6)	68 (11)	115 (18.7)	86 (14)	208 (33.8)
If the physician does mistakes in my treatment, I will tolerate them	122 (19.8)	70 (11.4)	113 (18.3)	67 (10.9)	243 (39.4)
If the physician does mistakes in my treatment, he will tell me about it	121 (19.6)	73 (11.9)	124 (20.1)	64 (10.4)	234 (38)
I sometimes worry that the physician may discuss some of my important personal information with others	55 (8.9)	40 (6.5)	112 (18.2)	67 (10.9)	342 (55.5)

The combined qualitative and quantitative studies have brought out the important dimensions of trust in physicians in a developing healthcare setting. It is evident that trust is a different concept in these resource-deprived settings compared to the settings in the developed world. This has important implications for trust as a concept, as well as understanding of trust in physicians in these settings.

References

Department of Statistics, G. of T. N. (2014). *Vital statistics*. Chennai. Retrieved from http://www.tn.gov.in/deptst/vitalstatistics.pdf.

Hall, M. A., Dugan, E., Zheng, B., & Mishra, A. K. (2001). Trust in physicians and medical institutions: What is it, can it be measured, and does it matter? *Milbank Quarterly, 79*(4), 613–639.

Mechanic, D. (1998). The functions and limitations of trust in the provision of medical care. *Journal of Health Politics, Policy and Law, 23*(4), 661–686.

Registrar General, I. (2011). *Census of India 2011: Provisional population totals-India data sheet*. Office of the Registrar General and Census Commissioner, India, Indian Census Bureau.

Chapter 4
What Influences Patients' Trust in Physicians?

Abstract This chapter describes the various factors that influence trust in physicians in the context of the study. Comfort and shared identity, personal involvement of the physician with the patient, communication, behaviours and approach of the physician, economic factors and health awareness of the patients, all have an influence on the trust in physicians. This chapter describes the frequencies of the responses of study participants on the various factors influencing trust in physicians. Further, it also reports the results of segmentation of the community after factor and cluster analyses methods into people who trust their physicians predominantly based on comfort, those who trust based on personal involvement of the physician, those who trust based on emotions and finally ones who trust based on behaviour and communication of the physician. It further describes the attributes of people within each cluster.

Keywords Trust in physicians · Determinants · Factors · Comfort · Personal involvement · Communication · Behaviours

After obtaining a conceptual understanding of what patients in developing healthcare settings perceived as trust in physicians, the next step was to understand what factors led to building of trust. I have already described in the previous chapter about how trust is defined as an attribute of perceived competence of the physician, assurance of treatment, willingness to accept limitations in the physician–patient relationship, respect and royalty. The factors determining trust are those which enhance these attributes, demonstrate them to the patients and thus promote the building of trust. The information about the various factors influencing trust in physicians was obtained from the qualitative interviews as well as the quantitative survey (Gopichandran & Chetlapalli, 2013a, 2013b; Gopichandran, Wouters, & Chetlapalli, 2015).

Comfort and Shared Identity

In the interviews that were conducted among people who had migrated from outside Chennai to the city for construction labour, I came across several a strong narrative of language and cultural barrier in approaching health care. These workers had migrated

V. Gopichandran, *Dynamics of Trust in Doctor-Patient Relationship in India*, SpringerBriefs in Ethics, https://doi.org/10.1007/978-981-15-0346-7_4

from various parts of the country and spoke different languages. However, the main language spoken in Chennai is Tamil, which many of these people did not know. So, they found it difficult to build a relationship with the local physicians. When they saw a physician who could speak Hindi, the language that many of these people spoke, they could immediately relate to this physician. They explained that if the physician understood and spoke their own language and also understood their culture and respected it, then it would increase their trust. This shared identity improves the physician–patient communication. Apart from this, it also gave them a sense of connectedness which greatly enhanced their trust. The idea of shared identity that came out in the interviews was largely revolving around the issue of language. It is also possible that shared identity in terms of place of origin, religion, caste, culture, could all bring about a similar influence on the trust in physicians.

In the interviews in the villages, one of the important determinants that emerged was the idea of comfort and ease with which the people could approach the physicians. A health facility that was close to home, which the people could access easily improved their level of trust. This is because there was a sense of 'this hospital is in my area and so it is mine' and this sense gave them a level of comfort which influenced their trust. Further, the familiarity of the hospital, its topography, the staff who work in the hospital and the procedures, gave them comfort which in turn influenced their trust.

It can be reasoned that the shared identity and comfort increased the chances of the patients to appreciate the various attributes of trust in the physician. Common language, culture and feeling of comfort helped the patients understand and evaluate the competence of the physician through easy communication. Easy access to the health facility and familiarity promoted a sense of assurance of treatment. Moreover, it is easier to respect and be loyal to someone whom the patient can understand and relate to culturally, than someone whom they cannot.

Personal Involvement of Physician with the Patient

People preferred when the physicians knew them personally and got involved with them at a personal level. They felt that this kind of a personal involvement increased trust. It is common for primary care physicians in rural areas and urban poor areas to know their patients personally. The patients invite the physician to their family functions. Physicians in turn attend these functions. Physicians even visit families which grieve the death of a loved one to show solidarity. In clinical settings, it is common to have discussions about the other members of the family and their lives. Such a personal involvement of the physician greatly enhanced the trust. The level of personal involvement starts from knowing the patient by name and addressing them by name. It extends to having personal conversations and discussions about family and life. People felt that such a deep personal involvement of the physician improved their trust in the physicians.

Patients described a trusted physician as someone who is like a member of their own family. In one interview, I spoke to an elderly woman who was living in the village alone. Her sons and daughters had migrated to the city. The traditional Indian families, especially the ones that the elderly is used to, are joint families where sons, daughters-in-law and grandchildren live with the parents and support them. However, the family dynamics are changing in this context and migration for work has become a common phenomenon. This has led to several elderly people in villages caught between the traditional expectations from the family and the modern changes in family dynamics. In this interview, the woman referred to the local physician as the surrogate for the son who had migrated out for work.

The other common practice in these settings is giving gifts to physicians. Many patients bring small gifts to physicians, fruits and eggs from their farm, sweets and dishes prepared at home. This is seen as a gesture of love and trust between the physician and patient. Personal involvement of the physician with the patient can be a very important means of demonstrating the genuine intent that the physician has towards the welfare of the patient. This demonstration of intent greatly enhances trust. If the physician is seen as someone who treats the patient like their own family, then it demonstrates to the patient that the physician would treat them like how she treats her own family.

It is important to note that such a factor influencing trust is unique to developing healthcare contexts. In studies from the Western cultures, professionalism, which is a clear distinction between personal and professional relationship and drawing clear boundaries in the physician–patient relationship, is described as an important attribute of a physician. Boundary crossing, especially in the context of romantic relationships between physician and patients has been described as a negative consequence of reduction of formalities in the physician–patient relationship. This is a risk that needs to be closely borne in mind.

Communication, Behaviour and Approach of the Physician

Previous scholars of trust in physicians have described that when patients asses the competence of their physicians, it is the communication competence that they are assessing rather than the technical competence. This is because there is an inherent differential in the level of knowledge about health and human body between the physician and the patient. Therefore, some have described communication and behaviour as dimensions of trust. During the initial stages of analysis of the qualitative interviews, we thought that communication and behaviour of the physician should be treated as a dimension of trust in physicians. However, we realized that the assessment of competence was based on experience, both personal as well as shared assessment with the community. This assessment was based on whether the patients got better with the treatments provided by the physician and not based on their communication. The patients were willing to trade-off the communication skills, politeness, kindness and other behavioural aspects if the outcome of treatment was good. Therefore, we decided to use communication, behaviour and approach of the physician as a determinant of trust rather than a dimension. Better communication and behaviours of the physician would help them demonstrate the core dimensions such as perceived competence and assurance of treatment. However, communication and behaviour were not of such core importance that the patients were willing to compromise on it if the outcomes of treatment were good.

The people who were interviewed mentioned that the physician should spend time, listen to their complaints and show understanding of their condition. After listening and understanding their problems, the physician should perform a thorough physical examination. In an overcrowded clinic in these settings, it is common for physicians to not even touch a patient and examine her before writing a prescription for medications. In this context, patients explained that spending time to perform a thorough physical examination increased their trust. The other aspect of communication that patients expected was for the physician to explain their condition and the treatment clearly. They also expressed that physician should give time to patients to ask questions and should clarify their doubts.

A young migrant construction labourer explained her experience of attending a doctor's clinic. In her narrative, I could clearly hear the lack of trust in physicians. She said, 'Once I had an insect bite in my leg. My leg swelled up and it was very painful. So I went to the local physician. The physician there just looked at me and wrote something in the prescription and sent me away. She did not speak to me. She did not even touch me (holds my hand and gestures). She did not even feel my pulse. I felt very upset. I will never go back to this physician.' This is an example of how talking, listening, touching, examining and explaining are all important determinants of trust in physicians. In this situation, there was a language barrier between the physician and the patient as they did not understand or speak the same language. In such situation, the only way the physician can demonstrate her genuine intent in making the patient feel better from her suffering and pain is to perform a thorough examination and attempt to communicate through body language. When all channels of communication fail, there is no way for the patient to asses whether to trust the physician or not. This has prevented the formation of trust in the relationship.

Economic Factors

In developing healthcare settings with resource limitations, no discourse on trust in physicians can be completed without a discussion of economics of health care. This theme was evident in most of the interviews. As seen previously, a physician's perceived competence was assessed based on whether they treated free of cost or for a fee. The higher the fee, the greater the perceived competence of the physician. In fact, it was found that patients were willing to pay a high cost for treatment, even at the expense of taking a heavy loan, if there was an assurance of good treatment. In settings where there is no universal health coverage with financial protections in place, healthcare expenditure is a serious concern. It often leads to catastrophic expenditure and pushes people into poverty and indebtedness. While on one hand, there is a belief that the expensive private practitioners are more competent than the government physicians who do not charge a fee, there is also a sense of doubt about the physicians who charge a high fee. One of the patients mentioned that he was doubtful whether the treatment costs the amount that he paid, or whether the doctor was exploiting him. In resource-poor settings, in the context of lack of access to universal health coverage, such doubts are common. It is important to note that such a doubt, while it may not completely negate the trust, it may keep the people sceptical about the physician. It would keep them in a situation of 'trust but verify' as described in the introductory section.

It is possible that if payment for care is removed from the physician–patient relationship, it could foster a greater level of trust. This is because it may remove the doubt regarding the appropriateness of the payment to the physician. With economic consideration removed from the dynamics of the physician–patient relationship, it is possible that the assessment of intentions of the physician to promote welfare of the patient might become clearer, thus fostering trust.

Health Awareness of People

In the interviews among physicians, one of the important themes that emerged was that patients are nowadays more aware of diseases, treatments and options. They attributed this to ease of access to information, brought about by the revolution of information and communication technology. They referred to this as the 'Google culture'. They said that patients are now looking up their symptoms and diseases on the Internet and asking very smart questions. The physicians felt that this access to information is dangerous as many times the information is wrong. So, they felt that such easy access to unreliable information only reduced the level of trust in the physicians.

A physician working in a government primary health centre mentioned that patients have become aware of their rights and entitlements in the health centre and therefore they have become highly demanding. He narrated an instance where he was called to the district headquarters for an emergency meeting and so had to leave the health facility urgently without seeing the patients. On seeing this, the patients waiting there raised their voice and staged a protest. So, he had to immediately abandon his visit to the headquarters and had to return to the facility to organize another doctor for the clinic. He expressed that nowadays he could not take patients for granted as the patients have become highly aware. Raised awareness levels among the people leads to a stage of 'trust but verify' which has been previously described in the west.

It is important to note that while trust in physicians can be positively influenced by personal involvement of the physician, providing comfort and shared identity, and good communication and behaviour, it can be reduced by expensive treatments and raised levels of community awareness. These findings have an important implication on whether trust is always a 'common good' to be aimed. Expensive treatments and lack of awareness make patients vulnerable. This subjects them to exploitation. In this context of vulnerability and exploitation, trust can be dangerous. Therefore, even though cost considerations and awareness may compromise the levels of trust, such a balance in trust is essential in the best interest of the patients to prevent undue exploitation.

Quantitative Analysis of Factors Determining Trust in Physicians

Based on this broad understanding of what influenced the trust in the physician–patient relationship, I developed some questions reflecting these determinants of trust and included them in the same quantitative survey that was described in the previous chapter. I added some items which were not obtained in the qualitative interviews such as attire of the physician and appearance of the physician. A total of 29 questions were included in the survey. To study if there are some patterns to what influenced the trust of the patients in their physicians, an exploratory factor analysis of the responses to these questions was performed. Exploratory factor analysis is a statistical method which evaluates the degree of correlation between the responses for the various questions and groups the questions for which the responses are strongly correlated. After this grouping is done, it is easy for the researcher to identify the common theme linking these questions and explain why the responses to these questions are strongly correlated. The assumption here is that there is a common underlying 'latent variable' or hidden concept which makes the responses to these questions to be strongly correlated. Based on the heuristics that are provided for exploratory factor analysis, I decided that the 29 questions can be grouped into 5 themes. These are shown in Table 4.1.

It is clear from the table that the respondents of the survey weighed communication and behaviour of the physician as the strongest determinant of trust. This was followed by the level of comfort provided by the physician and health facility. Personal involvement of the physician, appearance and shared identity were less important determinants. The exploratory factor analysis also very clearly grouped the questions into these meaningful groups with the common underlying themes as explained above.

One of the outputs of the exploratory factor analysis is the regression factor score. This factor score gives a score for each of the respondent in each of the themes identified based on their response. This score can be interpreted as the level of that 'latent variable'. For example, if the individual A obtained a score of 0.2345 in the factor 'communication and behaviour of the physician' compared to individual B who obtained a score of 0.6789, this can be interpreted as person B giving more importance to communication and behaviour as determinants of trust in physicians compared to person A. These factor scores were then used to perform a hierarchical cluster analysis. Cluster analysis is a statistical technique that groups individual

Table 4.1 Factors which influence trust in doctors

S. No	Latent factor	Questions	Respondents who agreed (%)	Factor loading
1	Communication and behavioural factors	The doctor explains to me clearly about the illness and its treatment	80.3	0.668
		The doctor listens to me patiently	79.8	0.666
		The doctor always has a smile on his/her face	71.5	0.658
		The doctor is a person who always talks the truth	75.3	0.633
		The doctor talks to me kindly	71.4	0.511
		The doctor treats everybody similarly irrespective of religion, caste, language or ability to pay, etc.	74.2	0.510
		The doctor does not discriminate against me for reasons of religion, caste, language etc.	61.4	0.449
		The doctor examines me with a stethoscope	71.8	0.434
2	Comfort	Feel comfortable talking to him/her	63	0.860
		I do not have any inhibitions with him/her	53.9	0.858
		It is easy for me to approach the doctor	56	0.779
3	Personal involvement	The doctor recognizes me and gives me special concessions	48.8	0.744
		The doctor knows my family situation	55.2	0.701

(continued)

Table 4.1 (continued)

S. No	Latent factor	Questions	Respondents who agreed (%)	Factor loading
		The doctor treats me like his/her own family member	57	0.661
		The doctor knows me by name	53.3	0.647
		The doctor tries to help me beyond just my medical needs	51	0.576
		The doctor accepts the gifts and presents that I give	31	0.549
		The doctor belongs to my village/town	51	0.467
		The doctor is a person who does not give much value to money	57.4	0.406
4	Simple and elegant appearance	The doctor wears white or light colours clothes	47.5	0.759
		The doctor appears simple	55.2	0.734
		The doctor wears clean clothes	66.4	0.722
		The doctor looks elder	52.5	0.566
5	Shared identity	The doctor belongs to my religion	23.5	0.751
		The doctor belongs to my caste/tribe	22.4	0.743
		The doctor understands my beliefs and practices	46.1	0.697
		The doctor speaks the same language as me	65.6	0.427

Table 4.2 Cluster centres based on factor scores of the five domains identified

S. No		Trust based on comfort $n = 67$	Trust based on personal involvement $n = 133$	Trust based on emotional factors $n = 276$	Trust based on communication and behaviour $n = 140$
1	Personal involvement of the physician	−0.35366	1.39398[a]	−0.58464[b]	−0.00245
2	Communication and behavioural competence	−0.62375[b]	−0.31679	−0.46689	1.51989[a]
3	Simple and elegant appearance	0.06329	−0.32628	−0.04605[a]	0.37046
4	Shared identity	0.00172	0.30827	−0.04379[a]	−0.20736[b]
5	Comfort	2.20438[a]	−0.10570[b]	−0.53013	0.09058

[a]Dimensions with maximum score
[b]Dimensions with least score

respondents to the survey into clusters based on relatedness of their scores. In other words, it will group the people who have similar scores. Therefore, while exploratory factor analysis reduces the questions in the survey into groups based on shared meaning of the questions, cluster analysis groups people based on the relatedness of the scores that they have obtained in the various latent variables. Hierarchical cluster analysis of the factors scores, using Ward linkage showed that the respondents can be optimally classified into four clusters. K means cluster analysis was done with the regression factor scores and convergence was achieved in 12 iterations. The participants separated into four clusters whose cluster centres are shown in Table 4.2.

It is seen that respondents who grouped in cluster 1 ($n = 67$) favoured ‘comfort’ and did not give importance to ‘communication and behavioural competence’ or ‘personal involvement of the doctor’. Cluster 2 ($n = 133$) members were interested in ‘personal involvement of the doctor’ but not in ‘comfort’, ‘simple appearance’ or ‘communication and behavioural competence’. Cluster 3 ($n = 276$) members gave importance to ‘simple and elegant appearance’ and ‘shared identity’ dimensions, whereas they were not much interested in ‘personal involvement of the doctor’ or ‘communication and behavioural competence’. Members of cluster 4 ($n = 140$) looked for ‘communication and behavioural competence’ element but did not care much for ‘shared identity’ or ‘comfort’. The clusters of the people thus obtained were labelled as those who trust based on ‘comfort’, ‘personal involvement’, ‘emotional factors’ and ‘communication and behaviour’. It was seen that in the setting where this

Table 4.3 Characteristics of the clusters of the community based on factors influencing their trust in physicians

Cluster name	Characteristics of the members
Comfort-based trust	Women (62.7%) Rural (86.6%) Older persons (53%) Lower education levels (73%) Homemakers, unemployed, labourer (65%)
Trust based on personal involvement	Younger persons (62.4%) Urban (40.6%) Higher education levels (70%) Professional, business, service (68%)
Emotional trust	Rural (83.3%) Lower education levels (70%)
Trust based on communication and behaviour	Women (70%) Urban (10.7%) Younger persons (78.6%)

study was conducted, maximum number of people placed their trust in physicians based on emotional factors such as shared language and identity. Lesser number of people belonged to the other clusters.

The characteristics of the respondents who grouped into each cluster were further studied. It was found that the people in the 'comfort-based trust' cluster were predominantly older, rural, women with lower education and lower levels of occupation (labourers, unemployed, etc.). Those who had 'personal trust' were of younger age, higher educational attainment and higher occupational levels. The segment of the community who had 'Emotional trust' was rural and had lower educational attainment. Those who had 'objectively assessed trust' seemed to be similar in profile with those who had 'personal trust' except the fact that they were predominantly younger rural women. People with personal trust were also more from urban residence compared to the other groups. These findings are shown in Table 4.3.

People who trust their physicians predominantly based on comfort levels were predominantly rural, women, older and belonging to lower educational and occupational attainment. This is a group who is rendered highly vulnerable by virtue of their socio-economic status in the society. It is possible that such high level of vulnerability leaves very little option than to depend on comfort of accessing the physician, approaching and consulting them. All other factors such as behavioural competence, personal involvement of the doctor, shared identity, appearance, etc., are secondary when compared to comfort of approaching the doctor. Social vulnerability leads to a lower threshold for trust, in this case, the mere comfort of access and approach.

Twenty-one per cent of the sample were clustered into the people who placed their trust on the physicians predominantly based on personal involvement. Younger people with higher levels of education and occupational attainment were grouped into this cluster. Compared to the comfort-based cluster, this cluster had a lesser level of vulnerability. Therefore, it was possible for this cluster to negotiate trust based on factors beyond access and approach. It seemed that at this higher level of social status, comfort and access was not just enough, but the people also preferred that the doctor gets involved with them personally.

Most of the respondents, about 45%, clustered in the group who trusted their physician based on emotional factors such as shared identity, common language and familiarity. They placed their trust based on common language, religion, caste and cultural beliefs. This was narrated as a sense of emotional connectedness. The shared language and cultural values apart from helping in communication, also helped in a sense of emotional bonding. This cluster again comprised of people belonging to rural background and having a lower educational attainment. This could be considered as a state of vulnerability. The level of vulnerability in this state again lowers the threshold of trust and operates it at the level of emotional connectedness offered by shared language and cultural values.

The fourth cluster had 22% of the respondents. This cluster placed their trust in the physicians based on communication and behaviour of the physician. This is the highest level of trust threshold that was found in this study. The cluster comprised predominantly of women from urban areas and who were younger.

Of these clusters those who trusted based on comfort and had an emotionally assessed trust seemed to belong to marginalized segments of the society. On the other hand, those who had personal trust seemed to belong to the higher socio-economic status. This study clearly demonstrates that with resource deprivation, the factors that influence trust in doctors tend to be in terms of emotional assessment rather than perceived quality of care or behavioural competence. The other explanation of this phenomenon could be that people from the better education and occupation background could articulate their expectations from the doctor–patient relationship clearly compared to those from the deprived groups.

Conceptual Model of Trust in Physicians

At this stage, I attempted a complete holistic statistical model of trust in physicians using the variable that I had in the study. I used a method called structural equation modelling for this analysis. Structural equation modelling is an advanced statistical modelling technique which uses a combination of correlations and regressions and represents it diagrammatically to depict a holistic picture of relationships of various observed variables to underlying latent variables. For example in this case, the structural equation modelling gave me the chance to depict 'trust in physician', the various dimensions of trust in physicians that I could measure using various items in the questionnaire, the various factors that influence trust in physicians along with various items measuring these factors influencing trust in physicians in a single model. The technique not only helped me represent them as a single model but also gave me the chance to see the magnitude of contribution of each component in explaining the concept of trust in physicians. This structural equation model is represented in Fig. 4.1.

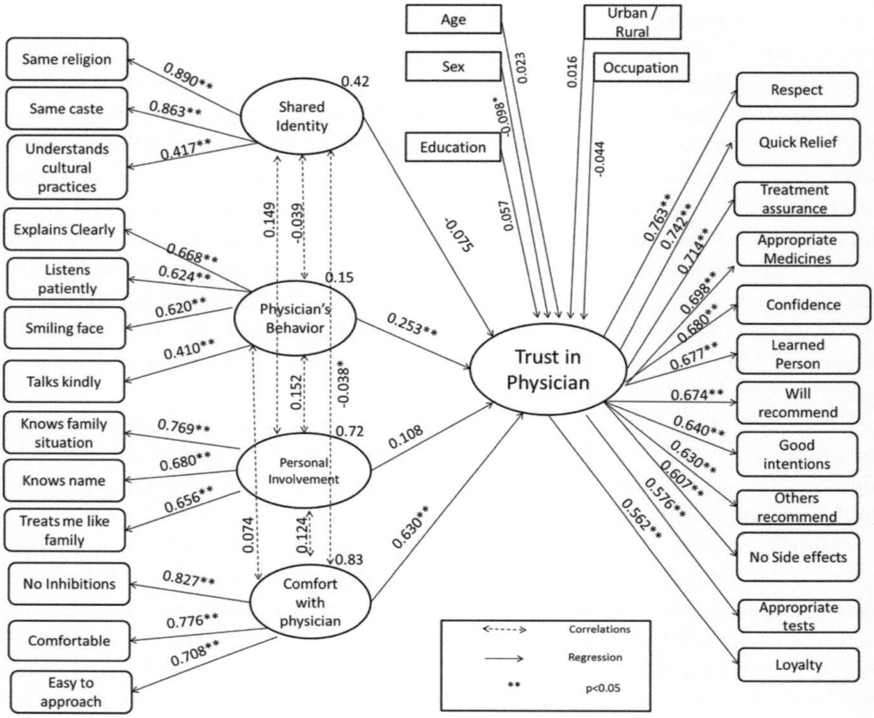

Fig. 4.1 Conceptual model of trust in physician

The model shows the regression coefficients of each of the 12 component statements' contribution to the idea of trust in physicians (right side of the figure). Sex of the patient ($\beta = -0.098$) had a significant influence on trust. Women had lesser trust in physicians compared to men. The other socio-demographic factors did not have a significant influence on trust in the physician. Shared identity ($\beta = -0.075$, $p > 0.05$) did not seem to influence trust. Physician behaviour ($\beta = 0.253$, $p < 0.05$), and comfort with the physician ($\beta = 0.630$, $p < 0.05$) influenced trust in the physicians. Table 4.4 shows the standardized regression coefficients, the bootstrap confidence intervals and p values for each association in the model. The intercorrelations between the four factors influencing trust in the physician are shown in Table 4.5. The correlation between behavioural factors and personal involvement was higher than the correlation between comfort and personal involvement. There is very low correlation between behavioural factors and comfort with the physician.

The final model had an SRMR of 0.044, RMSEA of 0.043 (90%CI 0.039–0.050), and CFI of 0.942 all indicating a good model fit. The Bollen–Stine chi-square p value was <0.01. But since this chi-square is sensitive to high sample size, this was not considered as a criterion for rejecting the model.

Table 4.4 Structural equation model of trust in physicians—standardized regression weight and bootstrap confidence intervals

Independent variable	Dependent variable	Standardized regression weights	Bootstrap CI		Bootstrap *p* value
			Lower limit	Upper limit	
Trust	Comfort	0.630	0.519	0.725	0.001
Trust	Emotional	−0.075	−0.157	0.017	0.134
Trust	Behavioural	0.253	0.139	0.368	0.001
Trust	Personal involvement	0.108	−0.005	0.213	0.061
Trust	Education	0.057	−0.033	0.155	0.197
Trust	Occupation	−0.044	−0.129	0.045	0.318
Trust	Age	0.023	−0.054	0.099	0.526
Trust	Sex	−0.098	−0.181	−0.017	0.021
Trust	Rural urban	0.016	−0.053	0.091	0.596
Same religion	Shared identity	0.890	0.833	0.950	0.001
Same caste	Shared identity	0.863	0.794	0.918	0.001
Understands beliefs and practices	Shared identity	0.417	0.344	0.486	0.001
Explains clearly	Behavioural	0.668	0.570	0.750	0.002
Listens patiently	Behavioural	0.624	0.522	0.714	0.001
Smiling face	Behavioural	0.620	0.523	0.703	0.001
Talks kindly	Behavioural	0.410	0.299	0.525	0.001
No inhibitions	Comfort	0.827	0.775	0.873	0.001
Comfortable	Comfort	0.776	0.703	0.837	0.001
Easy to approach	Comfort	0.708	0.639	0.771	0.001
Knows family situation	Personal involvement	0.769	0.701	0.824	0.001
Knows name	Personal involvement	0.680	0.601	0.748	0.002
Treats me like family	Personal involvement	0.656	0.580	0.730	0.001
Respect	Trust	0.763	0.708	0.810	0.001
Quick cure	Trust	0.742	0.670	0.795	0.001
Treatment assurance	Trust	0.714	0.641	0.777	0.001

(continued)

Table 4.4 (continued)

Independent variable	Dependent variable	Standardized regression weights	Bootstrap CI		Bootstrap *p* value
			Lower limit	Upper limit	
Gives appropriate medicines	Trust	0.698	0.616	0.756	0.002
Get confidence	Trust	0.680	0.605	0.740	0.001
Learned person	Trust	0.677	0.601	0.742	0.001
Will recommend this doctor	Trust	0.674	0.613	0.730	0.001
Intention of doctor is only my treatment	Trust	0.640	0.560	0.708	0.001
Others recommend the doctor	Trust	0.630	0.556	0.699	0.001
No side effects to medicines	Trust	0.607	0.527	0.675	0.001
Does appropriate tests	Trust	0.576	0.492	0.651	0.001
Loyalty	Trust	0.562	0.492	0.623	0.001

Table 4.5 Correlation between the factors influencing trust in physician

S. No	Factor influencing trust	Factor influencing trust	Correlation coefficient	*p* value
1	Shared identity	Behavioural	−0.039	0.005
2	Shared identity	Comfort	−0.038	0.183
3	Shared identity	Personal involvement	0.149	<0.001
4	Behavioural	Comfort	0.074	<0.001
5	Behavioural	Personal involvement	0.152	<0.001
6	Comfort	Personal involvement	0.124	0.002

This figure shows the structural equation model of trust in physicians. The 12 items on the right are the various dimensions of trust. The left side shows circular latent variables which are factors influencing trust. To their further left are the items reflecting these factors. On top are the observed variables which are likely to influence trust.

In summary, various factors play out to influence trust in physicians in developing healthcare settings. The qualitative study revealed that comfort, personal involvement, behaviour and communication of the physician, economic factors and level of health awareness influenced trust, some positively and some negatively. The quantitative survey helped segment the community in clusters based on the factors influencing the trust in physicians. It was found that largely there were four groups of people, those who trusted based on comfort, personal involvement, emotional factors and those who trusted based on communication and behaviour. There was a distinct socio-economic differential in the level of vulnerability of the people belonging to these clusters with people from more vulnerable groups trusting based on comfort and emotional factors and less vulnerable groups trusting based on objectively assessed factors like personal involvement and communication and behaviour of the physician. The holistic structural equation model helped statistically validate the various ideas studied so far.

References

Gopichandran, V., & Chetlapalli, S. K. (2013a). Dimensions and determinants of trust in health care in resource poor settings—A qualitative exploration. *PLoS ONE, 8*(7).

Gopichandran, V., & Chetlapalli, S. K. (2013b). Factors influencing trust in doctors: A community segmentation strategy for quality improvement in healthcare. *BMJ Open, 3*(12), e004115. Retrieved from http://www.pubmedcentral.nih.gov/articlerender.fcgi?artid=3855707&tool=pmcentrez&rendertype=abstract.

Gopichandran, V., Wouters, E., & Chetlapalli, S. K. (2015). Development and validation of a socioculturally competent trust in physician scale for a developing country setting. *BMJ Open*, 1–13. https://doi.org/10.1136/bmjopen-2014-007305.

Chapter 5
How to Measure Trust in Physicians?

Abstract In this chapter, the process of development and validation of a scale to quantify trust in physicians in the local healthcare setting is described. The scale development process has been informed by both the classical test and item response approaches. At the end of the process of scale development and validation, we identified a 12-item trust in physician scale. The chapter also presents evidence of further validation of the scale in which its ability to discriminate between those who trust and those who do not was clearly brought out. There is a need to use this scale more extensively and test its performance across different population groups.

Keywords Trust in physicians · Scale · Psychometrics · Classical test theory · Item response theory

Psychometrics is the science of measurement of abstract psychological constructs such as 'trust in physicians'. There are several reasons why abstract constructs such as 'trust in physicians' need to be measured. The measurement of trust can be of great instrumental value at the clinical level. As described previously, trust in physicians will improve the quality of the physician–patient relationship, enhance disclosure of sensitive information by the patients, improve adherence to the recommended treatment and also improve the chances of regular follow-up. A physician who understands the level of trust can work towards trust-building, can tailor treatment and counselling based on the level of trust and also anticipate treatment behaviours based on trust. At the population level, trust in physicians can be seen as an important measure of quality of healthcare services available to the people. It is also important to have valid instruments to measure trust for research purposes. As my research progressed, it became evident that trust in physicians is a very unique construct in developing healthcare settings. With this conceptual understanding of trust in physicians, came the opportunity to attempt to measure trust in physicians in a typical developing healthcare context using a scale developed uniquely in this context. In this chapter, I will describe the efforts in developing a scale to measure trust in physicians and validating it in the local context.

Several scales exist which measure trust in physicians. Most of these scales have been developed and validated in the Western and developed healthcare contexts. A brief summary of the available scales that measure trust in health care is presented

V. Gopichandran, *Dynamics of Trust in Doctor-Patient Relationship in India*, SpringerBriefs in Ethics, https://doi.org/10.1007/978-981-15-0346-7_5

in Table 5.1 (Dugan, Trachtenberg, & Hall, 2005; Egede & Ellis, 2008; Goold, Fessler, & Moyer, 2005; Goudge & Gilson, 2005; Hall, Camacho, & Dugan, 2002; Piette, Schillinger, Potter, & Heisler, n.d.; Rose, Peters, Shea, & Armstrong, 2004; Safran et al., 1998; Thompson et al., 2004).

The scales presented in the table are not just those which measure trust in physicians. They also represent scales which measure trust in the health system, trust in health insurers, etc. These are very different from the trust in physicians. However, most of the dimensions of trust across all these scales are common. Another important point to understand from these scales is that they are all designed and validated in the developed healthcare settings.

Among all the pre-existing scales, the one which is closest in relevance to the Indian context is the Anderson and Dedrick's Trust in Physician scale. It is a 11-item scale developed in the 1990s in the USA. The scale has items relating to competence, fidelity, confidentiality, honesty and a global trust in the physician. However, attempts to validate this scale in the Indian context showed that some items in the scale did not capture the attribute of trust accurately. For example, an item pertaining to obtaining second opinion from another doctor and an item covering the aspect of confidentiality did not describe the construct of trust in physicians well in the local context. So, the validation study indicated a need for a new trust in physician scale for the local context. More details of this study can be found in a publication from our team (Kalsingh, Veliah, & Gopichandran, 2017).

Before describing the steps undertaken for developing and validating the scale to measure trust in physicians, I will briefly describe the process of scale development. Let us begin by describing what a 'latent construct' is. A latent construct is an abstract attribute, phenomenon, trait or state. Examples of latent constructs include happiness, beauty, sadness, trust, faith, etc. A psychometric scale is a set of questions, which is asked to an individual. The responses to this scale provide an indirect measure of the level of the construct of interest in that person. The scale will provide a numerical value based on the responses, which can be converted to the actual level of the latent construct based on certain psychometric rules. There are several steps involved in the scale development process. A scale is usually a reflective measure, meaning that changes in the latent construct will be reflected as changes in the response to the questions in the scale. In case of trust, when there is a change in the level of trust in physicians, it will be reflected as change in level of respect for the physician, expression of loyalty to the physician, etc. The scale developed in this context was a reflective scale. The first step in scale development is identifying the basic theory of the latent construct. Based on this comprehensive theory, items are developed, representing various dimensions of the core construct. These items are statements, for which the participant responds either positively, negatively or neutrally. The responses are directly converted to numbers. Then, a mathematical or statistical relationship is drawn up between these numbers and the level of the theoretical construct. This process is called psychometric theory.

There are two major ways in which the actually measured numerical value of the scale is converted to the level of the latent construct using the psychometric principles, namely the classical test theory (CTT) and the item response theory (IRT). The CTT

Table 5.1 Psychometric properties of scales to measure trust in health care

S. No	Name of the scale	Authors	Domains	No. of items	Cronbach's alpha
1	Trust in primary care physician scale	Hall et al. (2002) Wake Forest University Team	Fidelity, honesty, competence, global	10	0.93
2	Trust in physician scale	Anderson and Dedrick (1990)	Fidelity, competence, honesty, confidentiality, global	11	0.85
3	Patient trust scale	Kao, Green, Zaslavsky, et al. (1998)	Fidelity, competence, confidentiality, global	10	0.93
4	Medical professions trust scale	Hall et al. (2002) Wake Forest University Team	Fidelity, competence, honesty, global	11	0.89
5	Health insurance organization trust scale	Zheng, Hall, Dugan, et al. (2002) Wake Forest University Team	Competence, honesty, confidentiality, global	11	0.92
6	Whole health system trust scale	Straten et al. (2002) Netherlands	Patient focus of providers, policies, provider's competence, quality of care, information supply and communication, quality of cooperation	36	0.74–0.88
7	Multidimensional trust in healthcare system scale	Egede and Ellis (2008)	Trust in providers, trust in payers and trust in institutions	17	0.89
8	Abbreviated measure of patient trust in physician, insurer and medical profession.	Dugan et al. (2005)	Trust in physician, trust in insurer and trust in medical profession	15 (5 questions in each domain)	0.87, 0.84, 0.77

(continued)

Table 5.1 (continued)

S. No	Name of the scale	Authors	Domains	No. of items	Cronbach's alpha
9	Measure of trust in health insurers	Goold et al. (2005)	Organizational trust, doctor patient trust, perceived vulnerability, satisfaction with care	13	0.95
10	Healthcare system distrust scale	Rose et al. (2004)	Fidelity, competence, confidentiality, honesty/informed consent	10	0.75

method assumes that True Score on a latent construct is the Observed Score on the scale plus an error term. The CTT states that over infinite administrations of the scale with each administration independent of the previous one, the error terms are random and therefore tend to cancel out each other. Therefore, over infinite administration, the true score on the latent construct tends to be equal to the observed score on the scale. Therefore, the CTT method uses the observed score on the scale to indicate the level of latent construct (Lin, Lin, & Russell, 2008).

On the other hand, the IRT method uses an exponential relationship between a hypothetical level of latent construct and the probability of getting a correct answer at each level. This exponential relationship offers the chance to calculate the difficulty level of each question, its ability to discriminate between those who have high levels of the construct and low levels as well as the influence of guessing the correct response to a question. The IRT method allows the computation of effectiveness of each item in capturing the construct of interest, level of information provided by each item, the effectiveness of the whole scale and the level of information provided by the scale. In this study, I used both these approaches in a complementary manner to develop and validate a scale to measure trust in physicians in a developing healthcare context.

The previous chapters describe the process of qualitative research to develop a comprehensive theory of trust in physicians in the developing healthcare context. From this theory, a total of 31 statements were developed representing trust in physicians. Of the thirty-one items that were originally included in the questionnaire, nine were dropped because of redundancy and poor conceptual representation of their respective domains. The items that had to be removed reflected the domain of 'willingness to accept drawbacks in the doctor–patient relationship' (4 items), 'competence' (1 item), 'assurance of treatment' (1 item), 'doubts about the competence of the doctor' (2 items) and the single item on 'confidentiality'. The remaining 22 items had good item to total correlation (>0.4), average interitem correlation (0.4–0.7) and a Cronbach's alpha of 0.928. The items, their item to total correlations and Cronbach's alpha if respective items are removed from the scale are shown in Table 5.2.

The properties of the item shown in Table 5.2 are all those based on CTT. It looks at how each item performs as a part of the overall scale. The item to total correlation shows how much each item score correlates to the overall score on the scale. A high level of correlation between item and total score indicates that the item contributes substantially to the overall score on the scale. Cronbach's alpha is a statistical measure of the internal consistency of the scale. If all the items of the scale provide information about the same latent construct, then the responses to these items will be internally consistent. Therefore, a high level of internal consistency indicates that all the items in the scale capture the same construct. This is a measure of reliability of the scale. The high value of Cronbach's alpha seen for this scale indicates that the scale is highly internally consistent, thus indicating that it is highly reliable.

In order to test the construct validity of the 22-item scale, a confirmatory factor analysis was performed. Confirmatory factor analysis is a latent variable modelling technique where all the 22 items are fit as explanatory variables to a single endogenous variable—trust in physicians. After this, the model fit parameters are tested to see if such a single-dimensional model is fit for the data. If the model fit parameters

Table 5.2 Classical test properties of the selected 22 items of the new trust in physician scale

S. No	Item	Response		Item to total correlation	Cronbach's alpha if item deleted
		Average	Std. dev.		
1	The doctor does appropriate blood tests and other tests to diagnose my disease	0.808	1.25	0.532	0.9246
2	The doctor gives appropriate medications for my disease	0.567	1.068	0.6197	0.9233
3	The doctor prescribes appropriate number of medicines based on the nature of the illness	0.685	1.133	0.6362	0.9229
4	The doctor prescribes more expensive medicines for serious illnesses	1.128	1.351	0.4064	0.9271
5	The doctor's treatment relieves the illness quickly	0.619	1.072	0.6845	0.9223
6	The illness gets relieved with just one visit, there is no need for repeat visits	0.87	1.225	0.6177	0.9231
7	There are no side effects to the medicines prescribed by the doctor	0.881	1.294	0.5894	0.9236
8	Friends, relatives and neighbours speak well about the treatment provided by the doctor	0.856	1.208	0.5813	0.9237
9	Friends, relatives and neighbours recommend me to go to the doctor	0.818	1.212	0.6139	0.9232
10	I get the confidence that all my illness will get alright when I go to the doctor	0.784	1.132	0.6605	0.9225
11	There is a big crowd in the clinic of the doctor	0.696	1.112	0.4691	0.9256
12	If I go to the doctor, I will surely get good treatment for my illness	0.596	1.076	0.675	0.9224

(continued)

Table 5.2 (continued)

S. No	Item	Response		Item to total correlation	Cronbach's alpha if item deleted
		Average	Std. dev.		
13	The doctor gives me good treatment irrespective of whether I have money to pay	1.174	1.466	0.4557	0.9265
14	The main intention of the doctor is to treat my illness and not anything else	0.846	1.24	0.6376	0.9227
15	Irrespective of what time of the day it is, whenever I go, I can get good treatment from the doctor	1.307	1.493	0.4915	0.9258
16	Whatever illness I have, I will go only to this doctor	1.205	1.488	0.5911	0.9237
17	Even if I go to another doctor, I will take the treatment only if this doctor approves it	1.588	1.611	0.4571	0.927
18	I will bring my family members only to this doctor	0.976	1.347	0.6204	0.923
19	I will recommend only this doctor to all those who ask me	0.958	1.321	0.6956	0.9215
20	I respect the doctor a lot	0.643	1.095	0.7204	0.9216
21	I think the doctor is a very learned person	0.76	1.181	0.6624	0.9224
22	I admire the doctor	1.17	1.444	0.6057	0.9233

indicate that the data fits a single-dimensional structure, then the scale can be said to be structurally valid, or can be said to have construct validity. The confirmatory factor analysis model for a single-dimensional structure—trust in physicians—revealed that the SRMR was 0.06, CFI of 0.958, TLI of 0.951 and RMSEA of 0.045 thus indicating an acceptable model fit for a unidimensional construct. The regression weights of the individual explanatory variables loading on to the dimension of trust are shown in Table 5.3.

After confirming the unidimensional construct validity, a special statistical method was adopted to study if there is local dependence between the items. Local dependence means that items are closely clustered together a common factor because responses to these items are strongly dependent on each other, and not just because

Table 5.3 Confirmatory factor analysis showing the contribution of the latent trait 'trust' to each explanatory variable

S. No	Explanatory variable	Regression weight	S.E.	*p* value
1	The doctor does appropriate blood tests and other tests to diagnose my disease	0.565	0.01	<0.001
2	The doctor gives appropriate medications for my disease	0.653	0.078	<0.001
3	The doctor prescribes appropriate number of medicines based on the nature of the illness	0.672	0.083	<0.001
4	The doctor prescribed more expensive medicines for serious illnesses	0.431	0.088	<0.001
5	The doctor's treatment relieves the illness quickly	0.732	0.081	<0.001
6	The illness gets relieved with just one visit, there is no need for repeat visits	0.652	0.089	<0.001
7	There are no side effects to the medicines prescribed by the doctor	0.613	0.092	<0.001
8	Friends, relatives and neighbours speak well about the treatment provided by the doctor	0.592	0.085	<0.001
9	Friends, relatives and neighbours recommend me to go to the doctor	0.639	0.087	<0.001
10	I get the confidence that all my illness will get alright when I go to the doctor	0.687	0.084	<0.001
11	There is a big crowd in the clinic of the doctor	0.493	0.075	<0.001
12	If I go to the doctor, I will surely get good treatment for my illness	0.731	0.081	<0.001
13	The doctor gives me good treatment irrespective of whether I have money to pay	0.436	0.096	<0.001
14	The main intention of the doctor is to treat my illness and not anything else	0.644	0.09	<0.001
15	Irrespective of what time of the day it is, whenever I go, I can get good treatment from the doctor	0.565	0.1	<0.001
16	Whatever illness I have, I will go only to this doctor	0.581	0.104	<0.001

(continued)

Table 5.3 (continued)

S. No	Explanatory variable	Regression weight	S.E.	*p* value
17	Even if I go to another doctor, I will take the treatment only if this doctor approves it	0.427	0.105	<0.001
18	I will bring my family members only to this doctor	0.613	0.095	<0.001
19	I will recommend only this doctor to all those who ask me	0.702	0.098	<0.001
20	I respect the doctor a lot	0.762	0.084	<0.001
21	I think the doctor is a very learned person	0.700	0.088	<0.001
22	I admire the doctor	0.603	0.102	<0.001

there is a common latent variable underlying them. For example, there is an item ‘do you feel comfortable talking to this physician’ and another item ‘do you talk to this physician about personal problems’. The assumption is that there is an underlying latent construct ‘trust in the physician’ leading them to group into the same dimension. However, these two items are also locally dependent because only if a person is comfortably talking to the physician will they be able to talk about personal problems with them. Therefore, such locally dependent items are evaluated and only one of the two items is retained in the scale. For this, there is a statistical method called LD-X^2 bifactor statistics. These statistics were computed and minimal local dependence was identified between three items measuring competence, two items measuring shared community opinions about the doctor and three items measuring assurance of treatment. But the LD X^2 was not very high and hence no items were deleted from the model at this stage. The LD-X^2 matrix is shown in Table 5.4.

After establishing that the twenty-two selected items measured a single dimension, ‘trust’ and that local dependence was minimal, the data was fit in a Samejima’s Graded Response Model (Drasgow & Parsons, 1983). The 1PL model revealed an AIC of 27,578.96, BIC of 27,972.63 and the −2log likelihood −13,700.48. For the 2PL model, the AIC was 27,327.17, BIC was 27,813.72 and −2log likelihood was −13,553.58. Based on the lower values of AIC, BIC and −2log likelihood for the 2PL model, it was accepted as the most appropriate model. The Likelihood Ratio Test value was 293.8 and for 21 degrees of freedom, the p value was <0.001. Table 5.5 shows the item parameters of the graded response model for the 22 items.

It is clearly seen that most of the questions have very high level of discrimination ($a > 1.70$). The polytomous responses ranging from ‘strongly disagree’ to ‘strongly agree’ seem to perform optimally at increasing levels of the trust ranging from −2.0 (low levels of trust) to +0.2 (moderate level of trust) without much overlap. Item-level statistics of significance shows that ‘appropriate medicines’, ‘relieves illness quickly’, ‘one visit is enough’, ‘confidence’, ‘will bring family members to the doctor’, ‘will recommend the doctor to others’, ‘respect’ and ‘admiration’ are statistically significant items ($p > 0.05$). Further details of the item response theory

Table 5.4 Bifactor local dependence matrix for the items in the new trust in physician scale

Item	Domains represented by the items	Marginal X^2	1	2	3	4	5	6	7	8	9	10	11	12	13	14	15	16	17	18	19	20	21
1	Competence 1	1.5																					
2	Competence 2	4.1	9																				
3	Competence 3	3.9	8.9	17.1																			
4	Competence 4	1	5.5	10.8	16.6																		
5	Competence 5	8.4	6.1	4.1	5.6	6.4																	
6	Competence 6	3.2	2.5	4.4	3.3	8.5	5.1																
7	Competence 7	2.2	8.5	4.2	8.6	8.6	7.7	5.8															
8	Competence 8	3.9	6.8	8.6	5.3	8.7	5.9	8.8	10.8														
9	Competence 9	5.7	5.9	5.5	4.9	12.7	2.9	6.6	13.5	20													
10	Competence 10	6.7	5.1	6.2	4.2	9	9.8	9.4	9.1	12.9	7.2												
11	Competence 11	3	10.5	6.7	7	5.8	8.9	4.1	8.8	3.6	8.4	7.9											

(continued)

Table 5.4 (continued)

Item	Domains represented by the items	Marginal X^2	1	2	3	4	5	6	7	8	9	10	11	12	13	14	15	16	17	18	19	20	21
12	Treatment assurance 1	10.5	3.9	6	3.2	12.5	4.1	4.8	8.2	7.4	7	10.7	6.9										
13	Treatment assurance 2	1	8.1	8	6.7	7.4	10.6	8.7	9.4	7	9.7	9.2	8.3	8.3									
14	Treatment assurance 3	8.3	5.2	6.1	7.8	10.2	7	12.5	6.9	8.7	6.9	9.6	7.4	9.5	15.3								
15	Treatment assurance 4	2.6	5.7	4.9	6.5	10.6	9.9	9.1	18	8.1	13.2	10	14.4	10.8	12.4	14.1							
16	Loyalty 1	4.8	8.8	9.9	11.6	8.4	7.6	11.3	8.7	15.5	15.9	10.9	10.3	8.6	8.5	6.6	20.3						
17	Loyalty 2	2.1	7.2	10.7	11.8	14.1	5.7	10.2	13.1	11.3	9.6	7.1	12.8	11.5	8.7	14.1	13.3	15.4					
18	Loyalty 3	5.1	5.8	9	6.6	16.4	9	4.1	8.4	10.2	10.8	11.2	12.6	10.7	7	6.8	16.4	12	11.4				
19	Loyalty 4	10.8	6.7	6.7	7.8	10.6	10.1	7.8	8.5	11.1	10.4	14.3	11	9.1	16.5	14.3	15.3	15	18.1	14.2			
20	Respect 1	16.1	5.4	5.5	3.1	11.1	6.6	8.3	10.4	9.8	7.6	10.1	10	5.9	11.1	14.6	14.4	10.1	14.6	9	9.1		
21	Respect 2	7.7	3.7	6.5	4.6	5.9	3.7	6.2	9.6	4.6	2.8	7.5	5.4	7.1	10.5	10.8	13.2	8.9	14.6	9.7	11	10.7	
22	Respect 4	3.4	4.6	8.5	4.7	8.6	4.3	6.4	8.9	5.3	11.1	10.7	9.4	5.9	6.3	10.5	11	6.3	15.6	7.2	7	13.8	7.9

Table 5.5 Item parameters of the unconstrained graded response model of the new trust in physician scale

S. No	Item	*A*	*b*1	*b*2	*b*3	*b*4	S-X^2	*p* value
1	Does laboratory tests for diagnosis	1.75	−2.05	−1.5	−1.01	−0.48	159.46	0.005
2	Appropriate medicines	2.54	−2.06	−1.5	−1.17	−0.68	93.67	0.140
3	Appropriate number of medicines	2.45	−1.9	−1.55	−0.96	−0.51	141.34	0.001
4	Expensive meds for serious illness	1.22	−2.17	−1.65	−0.76	0.01	222.19	0.001
5	Relieves illness quickly	3.02	−1.83	−1.41	−1.09	−0.51	90.28	0.273
6	One visit is enough	2.14	−1.81	−1.44	−0.87	−0.25	126.8	0.231
7	No side effects of medicines	2.09	−1.72	−1.35	−0.86	−0.35	142.27	0.049
8	Community speaks well about doctor	2.18	−1.82	−1.48	−0.87	−0.28	150.46	0.003
9	Community recommends the doctor	2.46	−1.66	−1.38	−0.93	−0.32	122.57	0.047
10	Confidence	2.55	−1.85	−1.48	−0.94	−0.29	120.27	0.081
11	Big crowd in the clinic	1.81	−2.28	−1.76	−1.05	−0.52	224.29	0.001
12	Assured of good treatment	3.43	−1.8	−1.32	−1	−0.59	102.69	0.037
13	Good treatment irrespective of ability to pay	1.24	−1.85	−1.4	−0.71	−0.15	219.84	0.001
14	Main intention is to treat my illness	2.6	−1.64	−1.27	−0.85	−0.35	137.47	0.025
15	Assured of treatment at any time of the day	1.45	−1.58	−1.11	−0.57	0.05	229.94	0.001
16	Whatever the illness will go only to this doctor	2.15	−1.33	−0.97	−0.53	−0.15	175.8	0.002
17	Will take other doctor's prescription only with permission of this doctor	1.36	−1.18	−0.89	−0.23	0.27	222.35	0.001
18	Will bring family members to this doctor	2.39	−1.46	−1.2	−0.69	−0.26	123.1	0.222

(continued)

Table 5.5 (continued)

S. No	Item	*A*	*b*1	*b*2	*b*3	*b*4	S-X^2	*p* value
19	Will recommend this doctors to others	3.12	−1.35	−1.1	−0.69	−0.25	118.13	0.117
20	Respect	4	−1.58	−1.3	−0.93	−0.49	72.37	0.466
21	Learned person	2.97	−1.62	−1.35	−0.9	−0.4	125.76	0.011
22	Admiration	2.14	−1.31	−1.09	−0.57	−0.06	141.81	0.095

a—discrimination parameter, $b1, b2, b3, b4$—difficulty parameter, S-X^2 item-level diagnostic statistic

Table 5.6 Optimally performing items selected after combined classical test and item response analyses for the new trust in physician scale

S. No	Questions
1	The doctor does appropriate tests to diagnose my disease
2	The doctor gives appropriate medications for my disease
3	The doctor's treatment relieves the illness quickly
4	There are no side effects to the medicines prescribed by the doctor
5	Friends, relatives and neighbours recommend me to go to the doctor
6	I get the confidence that all my illness will get alright when I go to the doctor
7	If I go to the doctor, I will surely get good treatment for my illness
8	The main intention of the doctor is to treat my illness and not anything else
9	Whatever illness I have, I will go only to this doctor first
10	I will recommend this doctor to all those who ask me
11	I respect the doctor a lot
12	I think the doctor is a very learned person

analysis, the item characteristic curves, item information functions and test information function are shown in a detailed publication of the validation exercise of this scale (Gopichandran, Wouters, & Chetlapalli, 2015).

Based on how representative the items are of their dimensions, findings from classical test theory, item information and item-level diagnostics 12 of the 22 questions were selected in the final scale. The items that were selected are shown in Table 5.6.

Thus, this study helped develop, validate, test the various item-specific attributes and assess the level of accuracy of each statement in predicting the level of trust in physicians. The various processes described are the scale development, reliability testing using Cronbach's alpha, construct validity using confirmatory factor analysis, testing of local dependence and item-level accuracy and precision.

To further validate this scale to assess its ability to classify people as those who trust and those who distrust their physicians, another small research study was conducted among 300 randomly selected households in Chennai city. Out of a total 300

Table 5.7 Characteristics of the respondents for predictive validation

Characteristic	Categories	Frequency (%)
Gender	Male	101 (54.9%)
	Female	123 (45.1%)
Education	Secondary school	3 (1.3%)
	High school	25 (11.1%)
	College	63 (28.1%)
	Post-graduate	105 (46.9%)
	Doctorate	28 (12.5%)
Trust category	Trusting the doctor	124 (55.4%)
	Distrusting the doctor	110 (44.6%)

sampled households, 234 participated in the study. Of these, 124 answered the questionnaire with respect to the physician they trusted the most and the remaining 110 answered the questionnaire regarding the physician they distrusted. The characteristics of the respondents are shown in Table 5.7.

The characteristics of the study respondents show that there was a male preponderance and most of them had a higher level of education. The Likert responses of each of the items were scored on a scale of 1 (strongly disagree) to 5 (strongly agree). These scores were averaged for each item. The average score for each item was compared between the trusting and distrusting groups. This is shown in Table 5.8. Independent *t*-test showed that the means of the trusting and distrusting groups were significantly different from each other. The group which trusted their doctor had significantly greater scores than the group which distrusted the doctor. Most of the mean scores of the items in the trusting group were between 'agree' and 'strongly agree' range, whereas in the distrusting group, it was between 'neither agree nor disagree' and 'disagree' range. The discriminant function model that was fit for this study explained 100% of the variance in the items, and the Wilk's lambda was 0.475. The chi-square was 160.9 and for 12 degrees of freedom the p value for model fit was <0.001 thus indicating a good fit of the discriminant function model. The standardized canonical discriminant function coefficients are shown in Table 5.9.

The group centroids for the trusting group were 0.940 and distrusting group was −1.166. The cut-off score of the discriminant function at which respondents can be classified as trusting and non-trusting is −0.113. Figure 5.1 shows that around a total scale score of 48 and discriminant function score of around 0, there is a distinct separation of trusting and distrusting individuals.

The accuracy of classification of the respondents into those who trust or distrust their physician based on the discriminant function revealed a sensitivity of 83.9% and a specificity of 83%. The classification accuracy was 83.5%. This further indicates that the 12-item trust in physician scale is valid in discriminating between trusting and non-trusting individuals with a fair degree of accuracy.

Table 5.8 Comparison of mean scores between the people who trust versus those who do not trust their physician

S. No	Items	Mean score (SD)		*P* value (independent *t*-test)
		Trusting the doctor	Distrusting the doctor	
1	The doctor does appropriate tests to diagnose my disease	4.31 (1.15)	3.20 (1.13)	<0.001
2	The doctor gives appropriate medications for my disease	4.59 (0.81)	3.30 (0.96)	<0.001
3	The doctor's treatment relieves the illness quickly	4.38 (0.98)	3.07 (0.96)	<0.001
4	There are no side effects to the medicines prescribed by the doctor	4.03 (1.28)	2.68 (1.11)	<0.001
5	Friends, relatives and neighbours recommend me to go to the doctor	4.12 (1.23)	3.61 (1.03)	0.001
6	I get the confidence that all my illness will get alright when I go to the doctor	4.28 (0.95)	3.09 (1.11)	<0.001
7	If I go to the doctor, I will surely get good treatment for my illness	4.57 (0.7)	3.15 (0.90)	<0.001
8	The main intention of the doctor is to treat my illness and not anything else	4.10 (1.24)	3.28 (1.11)	<0.001
9	Whatever illness I have, I will go only to this doctor first	3.35 (1.56)	2.57 (1.06)	<0.001
10	I will recommend this doctor to all those who ask me	3.97 (1.33)	2.92 (1.14)	<0.001
11	I respect the doctor a lot	4.39 (1.03)	3.46 (1.09)	<0.001
12	I think the doctor is a very learned person	4.27 (1.05)	3.54 (0.97)	<0.001

Table 5.9 Standardized canonical discriminant function coefficients of each item of the trust in physician questionnaire

S. No	Item	Standardized canonical discriminant function coefficient
1	The doctor does appropriate tests to diagnose my disease	0.039
2	The doctor gives appropriate medications for my disease	0.291
3	The doctor's treatment relieves the illness quickly	0.289
4	There are no side effects to the medicines prescribed by the doctor	0.241
5	Friends, relatives and neighbours recommend me to go to the doctor	−0.091
6	I get the confidence that all my illness will get alright when I go to the doctor	−0.118
7	If I go to the doctor, I will surely get good treatment for my illness	0.763
8	The main intention of the doctor is to treat my illness and not anything else	−0.017
9	Whatever illness I have, I will go only to this doctor first	−0.090
10	I will recommend this doctor to all those who ask me	−0.029
11	I respect the doctor a lot	−0.033
12	I think the doctor is a very learned person	−0.111

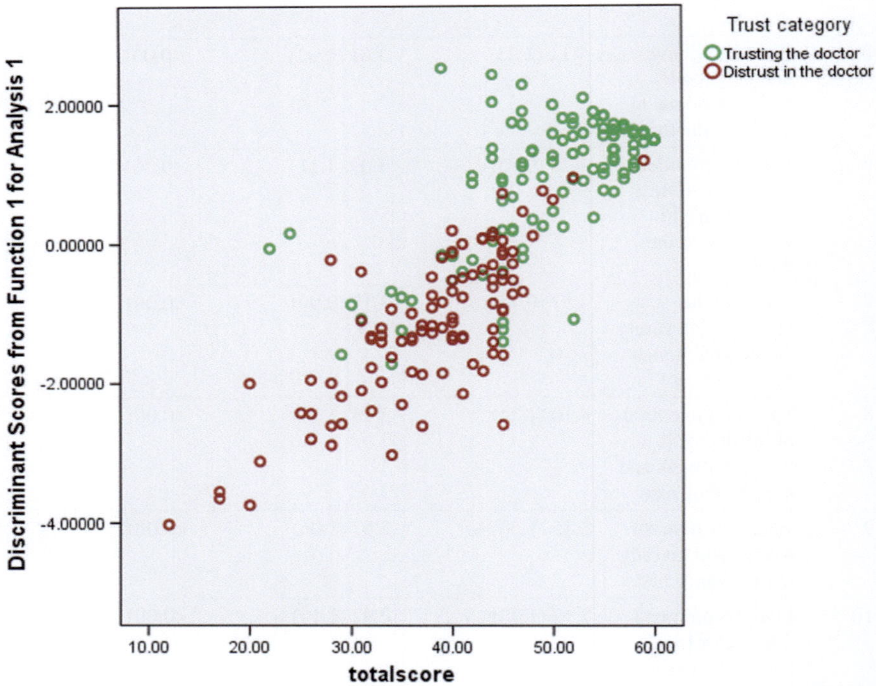

Fig. 5.1 Discrimination of trusting and non-trusting individuals by the trust in physician scale

In summary, this study helped in developing a valid and reliable scale to measure trust in physicians using the dimensions that are unique to developing healthcare settings. Further use of this scale and its application will add further weight to its validity.

References

Anderson, L. A., & Dedrick, R. F. (1990). Development of the Trust in Physician scale: A measure to assess interpersonal trust in patient-physician relationships. *Psychological reports*, *67*(3_suppl), 1091–1100.

Drasgow, F., & Parsons, C. K. (1983). Application of unidimensional item response theory models to multidimensional data. *Applied Psychological Measurement*, *7*(2), 189–199. https://doi.org/10.1177/014662168300700207.

Dugan, E., Trachtenberg, F., & Hall, M. A. (2005). Development of abbreviated measures to assess patient trust in a physician, a health insurer, and the medical profession. *BMC Health Services Research, 7*, 1–7. https://doi.org/10.1186/1472-6963-5-64.

Egede, L. E., & Ellis, C. (2008). Development and testing of the multidimensional trust in health care systems scale. *Journal of General Internal Medicine, 23*, 808–815. https://doi.org/10.1007/s11606-008-0613-1.

Goold, S. D., Fessler, D., & Moyer, C. A. (2005). A measure of trust in insurers. *Health Services Research, 41*, 58–78. https://doi.org/10.1111/j.1475-6773.2005.00456.x.

Gopichandran, V., Wouters, E., & Chetlapalli, S. K. (2015). Development and validation of a socioculturally competent trust in physician scale for a developing country setting. *BMJ Open,* 1–13. https://doi.org/10.1136/bmjopen-2014-007305.

Goudge, J., & Gilson, L. (2005). How can trust be investigated? Drawing lessons from past experience. *Social Science & Medicine, 61*, 1439–1451. https://doi.org/10.1016/j.socscimed.2004.11.071.

Hall, M. A., Camacho, F., & Dugan, E. (2002). Trust in the medical profession: Conceptual and measurement issues. *Health Services Research, 37*, 1419–1439.

Kalsingh, M. J., Veliah, G., & Gopichandran, V. (2017). Psychometric properties of the trust in physician scale in Tamil Nadu, India. *Journal of Family Medicine and Primary Care, 6*(1), 34.

Kao, A. C., et al. (1998). Patients' trust in their physicians effects of choice, continuity, and payment method. *Journal of General Internal Medicine, 13*(10), 681–686.

Lin, C., Lin, C., & Russell, E. M. (2008). Comparisons between classical test theory and item response theory in automated assembly of parallel test forms. *The Journal of Technology, Learning and Assessment, 6*(8).

Piette, J. D., Schillinger, D., Potter, M. B., & Heisler, M. (n.d.). Dimensions of patient-provider communication and diabetes self-care in an ethnically diverse population. *Journal of General Internal Medicine*.

Rose, A., Peters, N., Shea, J. A., & Armstrong, K. (2004.). Development and testing of the health care system. *Journal of General Internal Medicine, 19*(1), 57–63.

Safran, D. G., Taira, D. A., Rogers, W. H., Kosinski, M., Ware, J. E., & Tarlov, A. R. (1998). Linking primary care performance to outcomes of care. *Journal of Family Practice, 47,* 213–220.

Straten, G. F. M., Friele, R. D., & Groenewegen, P. P. (2002). Public trust in Dutch health care. *Social Science and Medicine, 55*(2), 227–234.

Thompson, H. S., Valdimarsdottir, H. B., Winkel, G., et al. (2004). The group-based medical mistrust scale: Psychometric properties and association with breast cancer screening. *Preventive Medicine, 38*, 209–218. https://doi.org/10.1016/j.ypmed.2003.09.041.

Zheng, B., et al. (2002). Development of a scale to measure patients' trust in health insurers'. *Health Services Research*, *37*(1), 185–200.

Chapter 6
Perceptions of Healthcare Providers on Trust in Physicians

Abstract Trust in physicians is perceived differently by the community compared to healthcare providers themselves. While the community laid more emphasis on communication, behaviour, comfort, personal involvement and competence, the healthcare providers emphasized on competence and minimizing errors are the priority. This reflected the difference in thinking about physician–patient relationships among physicians and the community. These aspects are described in this chapter.

Keywords Trust in physicians · Health care providers · Communication · Competence

My research mainly focused on community perceptions of trust in physicians. The preceding chapters describe these perceptions from a qualitative as well as quantitative perspective. At this point, I was left with an important question—'What do the physicians and other healthcare providers think about how patients trust their physicians?' In order to understand this, I developed another short survey among physicians and other healthcare providers including nurses, physiotherapists and others in the neighbouring state of Kerala which has health indicators and health system similar to Tamil Nadu.

Among 188 respondents, 34 (19.1%) were men, and 154 (81.9%) were women. One hundred and thirty of the respondents (69.5%) belong to the age group 20–30. The designation of the respondents includes 16 medical doctors, 31 dentists, 80 staff nurses, 16 paramedical staffs and 21 nursing clinical instructors. Among the respondents, 161 were working with private organizations, 25 with public institutions and other two were working in charitable organizations. This information is depicted in Table 6.1.

This study was done to assess the dimensions and determinants of trust in physicians from the perception of healthcare providers. Giving respect to the individual patient (90%) was reported as one of the most important determinants of trust from the provider's perspective. Optimal use of available resources (84%), co-operative workforce (83%) and appropriate referral services (82%) was also found to be significant factors that influence patient trust. Spending more time with the patient,

V. Gopichandran, *Dynamics of Trust in Doctor-Patient Relationship in India*, SpringerBriefs in Ethics, https://doi.org/10.1007/978-981-15-0346-7_6

Table 6.1 Characteristics of the healthcare providers

S. No	Characteristics	Frequency	Percentage
1	*Gender*		
	Male	34	19.1
	Female	154	81.9
2	*Age group*		
	20–30	130	69.14
	31–40	30	15.95
	41–50	17	9.04
	51–60	11	5.85
3	*Designation*		
	Doctors	16	8.51
	Dentist	31	16.48
	Paramedical	16	8.51
	Student nurse	9	4.78
	Staff nurse	80	42.55
	Head nurse	7	3.72
	Clinical instructor (Both dental and nursing)	21	11.17
	Interns	8	4.25
4	*Type of hospital*		
	Private	161	85.63
	Public	25	13.29
	Charitable	2	1.06
5	*Years of experience*		
	Fresher	33	17.55
	1–10	113	60.10
	11–20	19	10.10
	21–30	17	9.04
	31–40	6	3.19

exploring the experiences of the patient, performing more tests to diagnose illness and having more professional degrees were reported to be weak factors influencing trust (<50%).

The providers seem to give more importance to technical competence as a factor influencing trust followed by certain key behavioural competencies. They seem to give least importance to certain behavioural aspects like spending more time with the patient. This clearly points out that they are oriented more towards thinking from a disease perspective and from a technical viewpoint. There seems to be a gap in

Table 6.2 Healthcare providers' perceptions of factors influencing patients' trust in physicians

S. No	Statement	Proportion who "Agreed"
High contributors to trust in physicians (>80%)		
1	Understanding patient's problems and giving enough attention	97.9
2	Good-quality care	92.6
3	Attentive listening and helping them to express their difficulties	91.5
4	Giving proper respect to the patient and relatives	90.4
5	Being careful in dealing with the patient to avoid mistakes	86
6	Optimal utilization of available resources	84
7	Appropriate referrals	82.4
8	Co-operative workforce	80.3
Moderate contributors to trust in physicians (50–80%)		
1	Being gentle, friendly and approachable to the patient	76.6
2	Thorough knowledge and competence	78
3	Accurate diagnosis	75.7
4	Putting patient's health and well-being above all considerations such as cost	69.7
5	Regulatory mechanisms	66.5
6	Providing privacy while talking with the patient	61.2
7	Keeping the patient from being the victim of high cost	60.1
8	Providing correct information to patient about illness	58.5
9	Geographical proximity of healthcare facility to home	58
10	Participative decision making	55.9
11	Good reputation	50.5
12	Long-term relationship with the patient	50.5
Weak contributors to trust in physicians (<50%)		
1	Spending more time with the patient	46.3
2	Exploring patient's life experiences	49
3	Performing more laboratory tests	44.7
4	Declaring a lack of knowledge and limitation of expertise	33.5
5	More professional degrees	30
6	Confessing mistakes in treatment and apologizing	28

patient-oriented and patient-friendly thinking. The various factors influencing trust in physicians as reported by the healthcare providers are shown in Table 6.2.

The comparison of the perceptions of trust in physician between providers and the community is depicted in Fig. 6.1.

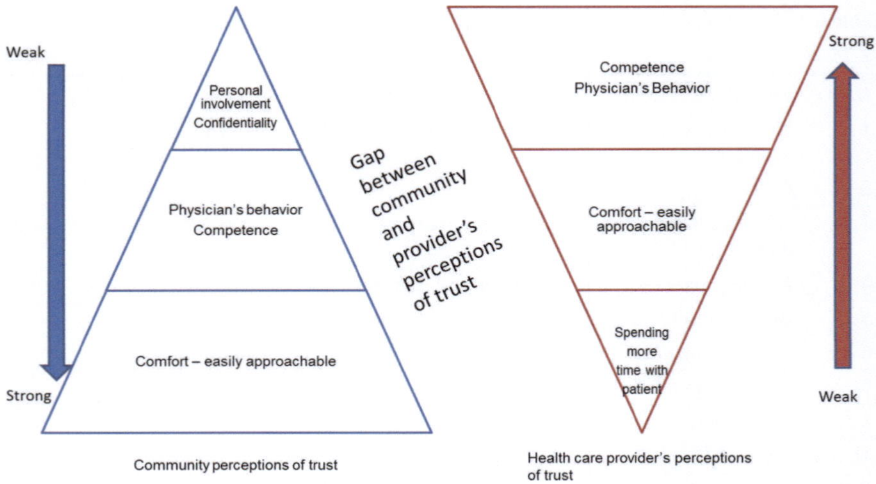

Fig. 6.1 Gap between the community and healthcare providers' perceptions of trust in physicians

It was clear from the previous structural equation analysis that the community preferred comfort and approachability of the physician as the main factor influencing trust. They laid moderate emphasis on behaviour of the physician and weakest emphasis on personal involvement of the physician. With respect to the dimensions of trust, they laid more emphasis on respect, loyalty and treatment assurance and less stress on competence. In this study, it was found that the physician's competence and behaviour were mentioned by the healthcare providers to be the strongest factors influencing trust. They laid lesser emphasis on comfort with the physician and least emphasis on certain behaviours like spending more time with the patient. Healthcare providers tend to approach the therapeutic relationship from an empowered plane and focus more on correct diagnosis, correct treatment and positive clinical outcomes. On the other hand, the community enters the therapeutic relationship in a vulnerable state with expectations of comfort and certain behaviours more than competence. Bridging this gap in perceptions would be an important step in strengthening the doctor–patient trust relationship. If providers can be empowered to address the social expectations of the community and the community empowered to participate more in competence judgments, it will strengthen the trusting physician–patient relationship.

A detailed qualitative study of perceptions of trust in the physician-patient relationship in India revealed important findings similar to the findings of this research (Kane & Calnan, 2017; Kane, Calnan, & Radkar, 2015). The providers also felt the need for a family physician model. They felt that such a model enhanced trust in the physician. Privatization and marketization of the practice of medicine have greatly influenced the public perceptions of trust in physicians. The respondents in that study also felt that the patients now have a challenging time as the trustworthiness of the physicians is determined by market forces. Many providers in that study also mentioned the importance of 'cut practice' and corruption in medicine as a major deterrent of trusting physician–patient relationships. The providers also felt that trust is shifting gradually from the person to technology. Patients now trust sophisticated equipment and advanced medical technology more than the physicians. The provider's expressed growing mistrust in the regulation of the practice of medicine. The various legal provisions and acts that governed the medical practice such as the Pre Conception Pre Natal Diagnostic Tests Act (PCPNDT Act) were perceived as unnecessary harassment by the regulators. The physicians felt that such regulations of their practice worsened the patients' trust in the physicians as well as the health system.

There was also a perception that trust between the modern medicine and other systems of medicine was declining. This was perceived to be related to extensive illegal cross practice. Physicians from alternative systems of medicine practicing modern medicine and presenting themselves as modern medical practitioners were viewed as a major factor impeding trust in the physicians.

With changing topography of medical practice in developing healthcare settings, the physician is decreasingly becoming the point of contact for patients while seeking health care. As seen earlier in the introduction chapter, trust in the health system is gradually moving away from an interpersonal trust in the physician to a systemic trust. There are multiple players, working together orchestrating the health system. Therefore, not only trust in the health system distributed across these players, trust in physicians has become a smaller component of the overall trust in the health system. While the community perceives the need for an easily approachable physician—the family physician or general practitioner—the providers felt that it is important to have competent and technically savvy physicians at the helm of affairs. This dichotomy is very prominent and indicates the need for a middle ground where both the physician's perceptions of trust and the community perceptions can meet. If this dichotomy continues unbridged, it would only lead to worsening the state of growing distrust in physicians.

References

Kane, S., & Calnan, M. (2017). Erosion of trust in the medical profession in India: Time for doctors to act. *International Journal of Health Policy and Management, 6*(1), 5–8. https://doi.org/10.15171/ijhpm.2016.143.

Kane, S., Calnan, M., & Radkar, A. (2015). Trust and trust relations from the providers' perspective: The case of the healthcare system in India. *Indian Journal of Medical Ethics, 12*(3), 157–168.

Chapter 7
Trust, Trustworthiness and the Physician–Patient Relationship in Developing Healthcare Settings

Abstract Trustworthiness is an important virtue in physicians. It is the underlying virtue which makes patients trust their physician. The social construct of trust in physicians is valuable only if it is based on the trustworthiness of the physician. There are situations where trust is placed on the untrustworthiness and vice versa. The basis of trustworthiness is the ethical practice of medicine and consistent demonstration of trustworthy behaviours by the physician. This chapter highlights the role of trustworthiness of the physician and the role of each physician as an exemplar and role model of trustworthiness for the profession to flourish.

Keywords Trust in physicians · Trustworthiness · Physician-patient relationship · Virtue ethics

Trust in the physician is of intrinsic and instrumental value in health care. It is of instrumental value because a trusting physician–patient relationship can lead to a sense of well-being, better physician–patient clinical encounters, more meaningful engagement in treatment and adherence to treatment instructions leading to better clinical outcomes (Gilson, 2003; Illingworth, 2002; Lee & Lin, 2009; Zaner, 1991). The intrinsic value of trust in the physician is more abstract and difficult to define. Having a trusting relationship with a physician has intrinsic value because it reflects the trustworthiness of the physician. In this chapter, I will describe the intrinsic value of trust in physicians, the foundations of trustworthiness of a physician and the role of ethical practice of medicine in trust and trustworthiness.

Societies function because of trust. Relationship between any two members of the society is bound by trust. If trust were to be lacking, it is very difficult for any societal transactions to take place. In the absence of trust, there is a need for strict regulation of all social activities. This is true for most social functions including buying goods and products from the market, availing services such as the Internet, mobile phone services and health care. This need for trust, for ease of social function and more specifically the healthcare service delivery provides it an intrinsic value in health care. If I trust my physician, I do not need to assess, evaluate or test the physician each time I have a health problem before I seek her care. By taking the uncertainty of quality of attention and care that I will receive from the physician, away from the equation, trust offers me intrinsic comfort and value (Saunders, 2017).

V. Gopichandran, *Dynamics of Trust in Doctor-Patient Relationship in India*,
SpringerBriefs in Ethics, https://doi.org/10.1007/978-981-15-0346-7_7

The point at which a person seeks health care is the point when she is most vulnerable either due to her own sickness or sickness of a near and dear one. This is a highly stressful period filled with uncertainty. At such a point, having a physician whom she trusts is a matter of great intrinsic value.

If I were to ask the simple question, 'Why is it good to have a physician whom I trust?', I can answer 'Because that way, I do not have to always test, evaluate or doubt his/her treatment'. Then I can further ask, 'Why is it good to have a physician whom I do not have to test, evaluate or doubt?', for which I can answer, 'That takes away the uncertainty at the time of my sickness and vulnerability'. Then I can go a step further and ask, 'Why should I take away uncertainty at the time of sickness and vulnerability?' and for that, the only answer can be 'Because that is the right thing to do, to reduce suffering'. Such a value, which comes to a point where there are no further reasons why such a phenomenon is good, is known as an intrinsic value. Therefore, having a physician whom I can trust is of intrinsic value (Saunders, 2017).

An important attribute of a physician is his or her trustworthiness. It is interesting that trust and trustworthiness do not always go hand in hand. There are instances where there is trust placed on those who are not trustworthy, and trust not being placed on those who are trustworthy (Goold, 2002; Saunders, 2017). For example, a physician may be highly technically competent, honest, sincere and trustworthy, but because of his/her inaccessibility and poor communication skills may not be trusted. A very average physician, who by virtue of poor clinical skills and poor moral integrity is not trustworthy, may be highly trusted because of perceived competence by the community and good communication skills. Understanding dimensions of trustworthiness is as important as understanding trust in the physician–patient relationships but is often challenging.

Trustworthiness is a virtue. A person's trustworthiness is assessed by his or her moral character including truthfulness, honesty, integrity and reliability. Virtues are defined by the attributes of people who are considered virtuous. It is a matter of role modelling. A physician who is trustworthy is a role model, and all the attributes of such a physician contribute towards the definition of trustworthiness. Virtue ethics define right and wrong based on the flourishing of human excellence. The question of trustworthiness is not about what a trustworthy physician should do, rather it is about how a trustworthy physician should be? Trustworthiness is less about doing and more about being.

A trustworthy physician is one who can be relied upon to always act in such a manner that will be exemplary. Let us take the example of a patient who has an acute abdominal pain. The physician examines the patient, does a few tests and recommends surgery. Taking the decision of proceeding with the surgery is a tough one. Given the commercialization of medical practice, the patient is now in a doubt whether the surgeon is unnecessarily recommending a surgery for commercial reasons. An appendix surgery will not do excessive harm to the patient; at the same time, the surgeon can make quick money. So, the patient is pushed into a trust conundrum. It is in this context that trustworthiness as a virtue is highly valuable. A trustworthy surgeon is an exemplar. He or she will always act in ways that flourish humanity.

They will always act honestly with integrity and reliability. However, such a trustworthiness is not up for scrutiny by the patient before going in for the surgery. This is the challenge of the concept of trustworthiness. While trustworthiness is the ideal virtue to have physicians and surgeons stand up to, it is almost impossible to assess. One way to assess trustworthiness in a physician is to evaluate consistent trustworthy behaviour. However, trustworthy behaviours may sometimes lead to erosion of trust. For example, a physician injects the wrong drug to a patient. Fortunately, the patient does not suffer from any serious adverse reaction. But the physician, who in keeping with the trustworthy behaviour of transparency, declares this mistake to the patient and apologizes for the error. But this declaration of the error erodes the trust in the physician. It is quite possible that lack of any negative outcomes in this case might help the patients appreciate the transparency of the physician, and this might build trust. However, if a negative outcome had resulted, it is not clear whether trust would have built.

Trustworthiness as a virtue is completely left to self-regulation by the physicians. It is completely left to the physicians to ensure that they are trustworthy. Neither is this trustworthiness possible to assess objectively nor is it clear to define independent of showing up other exemplary trustworthy physicians as examples. While trustworthiness is aspirational and important to build in physicians, it is not clear whether trustworthiness alone would do the job of building trust in physicians. Trust in physicians seems to emerge from a delicate balance between the moral character of the physician, his or her trustworthiness and the social construction of trust in the physician, which is more objectively evaluated by the people as described in the previous chapters. This means that while the social construct of trust in physicians may vary across different settings, between the developed and developing healthcare settings, trustworthiness as a virtue has very little variability, being the constant that defines the idea of trust in physicians.

One way of conceptualizing trustworthiness is practicing clinical medicine ethically and upholding the various principles of ethical practice of medicine. I would like to consider the important ethical considerations and see how holding these ethical considerations as the normative basis of the practice of medicine can uphold trust. I would also like to describe how sometimes the construct of trust and ethical practice of medicine can be in conflict.

Respecting the autonomy of the patients is one of the important ethical principles. Each individual is a moral agent who has the right to self-determination of the tests, treatments and procedures that can be done on him or her. Therefore, in respect with such a self-determination, the individual must be treated as an autonomous agent. In the most recent times, this has led to emergence of shared decision making in patient care (Shay & Lafata, 2015). This means the physician explains the choices of treatment available to the patient, discusses the pros and cons of the various choices and has a transparent talk in order to arrive at a shared decision. In this model, there is a need for active engagement of the patient with the treatment. However, sometimes high levels of trust in the physician preclude the engagement with the treatment options and discussions. The trust leads the patient to believe that whatever the physician does is in the best interest of the patient and disengages them from active participation in the decision. This is seen more often in developing healthcare settings. In these settings, where a level of paternalism is also intrinsic in the physician–patient relationship, such disengagement with treatment decision is a risk posed by high levels of trust. This disengagement with clinical decisions also throws the patients into a position of vulnerability to exploitation. Thus, compromise in autonomy that takes place because of high levels of trust predisposes the patients to exploitation. For example, the patient who has an abdominal pain and is asked to undergo surgery, if out of blind trust on the surgeon, refuses to engage in shared decision making, may be vulnerable to being subject to unnecessary surgery. Therefore, trust as a social construct, to be of intrinsic value, must be associated with trustworthiness as a virtue and consistent trustworthy behaviours of the physician.

The conflicting relationship between honesty, transparency and trust has already been described. If a physician, in the spirit of honesty and transparency discloses his or her errors to the patients, it may lead to erosion of trust. However, this does not mean that honesty and transparency are not of value. This only shows how trust in itself cannot be held as an independent value, in the absence of trustworthiness of the physician.

The balance between beneficence, non-maleficence and trust in the physician is also very delicate. In an example of a patient who suffers from a terminal form of cancer, if the physician withholds the diagnosis from the patient on request from the family, as they fear that it will push the patient into depression and worsen her quality of life and if the patient gets to know her diagnosis, it will erode the trust in the physician. The erosion of trust is acute if the vulnerability is more, as in the case of a terminal cancer. This balance is also difficult to maintain.

It can be argued that there are various degrees of rationality associated with trust in physicians. At one level, it is strategically the most convenient if the physician is trusted as it would be the option of least resistance. However, this strategic trust is subject to verification and nullification if experiences prove the contrary. At another level, trust may emerge from face value, from the approach and behaviour of the physician, prior to the treatment experience. This again is subject to scrutiny based on the treatment experience. At a completely different level is irrational trust, which is not subject to scrutiny. This is the kind of blind trust that children have on their parents. Children do not trust their parents based on their experiences with the parents. Their trust on their parents is not subject to scrutiny and verification. However, the trust is unquestioning, and it is inherent. This trust exists despite evidence to the contrary. Such a blind irrational trust can be dangerous, as it predisposes the patient to an extent of vulnerability. Such a blind and irrational trust cannot be justified in health care even under the best of circumstances, as greed and self-centeredness are as much human characteristics as generosity and altruism.

In summary, trust and trustworthiness seem to be valuable aspects of a good physician–patient relationship based on virtue ethics principles. A trustworthy physician is a virtuous physician, who is an exemplary role model of how a physician should be. Such trustworthiness is not possible to scrutinize and evaluate barring the exception of a few demonstrable trustworthy behaviours. There is also a subtle but definite difference between trustworthiness and trust. All trustworthy physicians are not necessarily trusted and vice versa. While it is desirable to have highly trustworthy physician, it is impractical to incorporate trustworthiness as a construct to evaluate physicians on. Trust can be articulated as an ethical principle that physicians should uphold on the basis of exemplary trustworthiness.

References

Gilson, L. (2003). Trust and the development of health care as a social institution. *Social Science & Medicine, 56*(7), 1453–1468.

Goold, S. D. (2002). Trust, distrust and trustworthiness: Lessons from the field. *Journal of General Internal Medicine, 17*(1), 79.

Illingworth, P. (2002). Trust: The scarcest of medical resources. *The Journal of Medicine and Philosophy, 27*(1), 31–46.

Lee, Y.-Y., & Lin, J. L. (2009). Trust but verify: The interactive effects of trust and autonomy preferences on health outcomes. *Health Care Analysis, 17*(3), 244–260.

Saunders, J. (2017). Trust and mistrust between patients and doctors. In *Handbook of the philosophy of medicine* (pp. 487–502).

Shay, L. A., & Lafata, J. E. (2015). Where is the evidence? A systematic review of shared decision making and patient outcomes. *Medical Decision Making, 35,* 114–132. https://doi.org/10.1177/0272989X14551638.

Zaner, R. M. (1991). The phenomenon of trust and the patient-physician relationship. In *Ethics, trust, and the professions: Philosophical and cultural aspects* (pp. 45–65).

Chapter 8
Trust in Physicians as a Quality Improvement Measure

Abstract Patient experiences and patient-reported outcomes are increasingly used to measure quality of health care. Patient satisfaction surveys have been used as a health quality metric for several years. It is possible to conceptualize trust in physicians as a quality indicator, complementary to patient satisfaction. Previous empirical studies have failed to demonstrate any substantial improvements in trust in health care following various interventions. These experiences need to be carefully evaluated while developing trust as a quality metric in health care.

Keywords Trust in physicians · Indicator · Quality of health care

This study has led to a conceptual understanding of trust in physicians in a developing healthcare setting. Further, it has also led to development and validation of a scale to measure trust in physicians in these settings. What next? How is this conceptual understanding of trust in physicians important? How can this scale be utilized to better understand the physician–patient relationship dynamics in the health systems? These are important practical questions that need to be addressed at the end of this study. In this chapter, I will describe the role of trust in physicians in the health systems of developing healthcare settings. I will describe how this trust is influenced by public trust in the system, and trust in other healthcare providers. I will further attempt to describe how trust in physicians reflects quality of not only the physician–patient relationship, but the overall quality of the health system performance. From these discussions, I will try to argue for the use of trust measurement to indicate quality of the health system. I will also reflect on the possibility of trust building exercises and their impact on improving quality of the health system.

The physician–patient relationship is a dynamic one set within the context of the health system. It cannot be viewed independent of the health system and its various components that are connected through a complex adaptive relationship. Trust in the physician is influenced by trust in the health system and vice versa. The same physician may be trusted when she practices in private and not trusted when she is working in a government hospital because of the facilities, infrastructure, cost and other factors. It is also is common to find a gross deficit in trust in the government health system, but some physicians within the system may continue to enjoy the trust of the community. It is also common to find some government hospitals trusted

V. Gopichandran, *Dynamics of Trust in Doctor-Patient Relationship in India*, SpringerBriefs in Ethics, https://doi.org/10.1007/978-981-15-0346-7_8

highly for certain healthcare services despite a gross mistrust in the government health system. For example, a government health facility in India is trusted highly for its hand surgery and microsurgery treatments despite the government health system itself not enjoying high levels of public trust. These examples indicate that there is a delicate dynamic between trust in physicians and trust in health system that can sometimes be concordant and at other times discordant.

Trust in individual physicians and healthcare providers within the system, the interpersonal trust, will ultimately contribute to the health system trust. But that alone is not enough most of the time. In addition to interpersonal trust, there is a need for a trust in the health system with respect to the other components of the health system. If the people have mistrust on the health policy, the health financing mechanism, availability of drugs and devices and quality of the health care, they may not trust the system purely based on the trust they have on physicians and other healthcare providers working in the system. This is seen in the former example of people trusting the same physician when she practices in private compared to her practice within the government system. Therefore, while trust in physicians is a necessary component of health system trust, it is not enough. It must operate in the context of trust in the other components of the health system.

Even within the various interpersonal trust relationships within the system, trust in physician is influenced by trust in other care providers within the system. Trust in the nurses, the technicians who perform tests and imaging services, the physiotherapist, the laboratory technician, all these trusting relationships influence trust in the physician. For example, in a hospital, if a laboratory technician performs a blood glucose test, looks at a value of 200 mg/dl of post-prandial sugar and tells the patient that she has diabetes and will be started on treatment, whereas the physician in her best clinical judgment wants to give the patient the benefit of doubt, defers starting any medications and starts only lifestyle modifications, this can lead to mistrust. A relationship of mistrust between the physician and other members of the team can also lead to substantial discomfort among the patients and lead to erosion of overall trust in the health system (Trachtenberg, Dugan, & Hall, 2005).

From the above description, trust in physicians may not on its own indicate quality of the health care provided by the health system. However, in a comprehensive theoretical understanding of healthcare quality, trust may play an important role. The Donabedian model of healthcare quality which includes structures, processes and outcomes serves as a good starting point to understand the complex construct of healthcare quality and the role of trust in physicians (Donabedian, 1988). According to this model, the quality of a health system is indicated by the various structures in it such as the physical infrastructure, personnel and other resources including staff training, capacity and financing mechanisms. It is indicated by the various processes which make up the healthcare delivery including diagnosis, treatment, preventive strategies, health promotion and health education. Finally, the quality is also indicated by outcomes including changes in health status, quality of life and behaviours. Trust in physicians will play an important role in the interface between the processes and outcomes.

Previous studies have shown that trust in physicians improves the process of history taking and declaration of sensitive information in the history. This improves the diagnosis. Trust in physicians also improves adherence to treatment and follow up, thus translating to good clinical outcomes. In addition to these two benefits, studies have also shown that trust in physicians reduces the need for second opinions and laboratory tests and imaging. This further cuts down healthcare cost. Trust in the physician has also shown to improve utilization of preventive and health-promotion services prescribed by the physician (Bonds et al., 2004; Lee & Lin, 2009; Mohseni & Lindstrom, 2007; Ohya et al., 2001; Piette, Heisler, Krein, & Kerr, 2005; Thom, Hall, & Pawlson, 2004). Given these substantial interactions between the process and outcomes of the health system, trust in the physicians is an important indicator of quality of care.

Existing evaluations of health system quality include patient satisfaction through exit interviews. There is an increasing emphasis on need to incorporate patient reported outcomes such as patient experiences and satisfaction with services in measurement of quality of care. Patient satisfaction is defined as the extent to which the patient is satisfied with the care provided to them in the health system. This includes their satisfaction with the physician. Satisfaction is an experiential measure which happens after the clinical encounter (Balkrishnan, 2003). With increasing importance given to patient-centred care, measurement of patient satisfaction has influenced quality of care in hospital and health systems. Interventions based on measures of patient satisfaction have led to improvement in quality.

Trust in physicians is a forward-looking construct, and in a way, it can be said to be complementary to the construct of patient satisfaction. It is the optimistic willingness to accept the vulnerability and subject oneself to the physician in the anticipation that the physician will do what is in the patients' best interest. Thus, it can also be effectively leveraged to indicate a dimension of quality of health care.

Empirical studies have shown challenges in trust building as a quality improvement strategy. Thom et al., gave training to doctors on certain behaviours which were associated with increased patient trust. The training focused on clearly addressing the patients' concerns, thorough history taking and physical examination, involving the patient in decision making, addressing the doubts and questions that the patients have and arranging for follow-up care. There was only a two-point difference in the post-trial physician trust scores between the intervention and control groups. There was no statistically significant difference between the two groups (Thom, 2000). Hall et al., studied whether disclosure of the information about the incentives that doctor received for complying with managed care protocols increased the trust of patients on their doctors. He showed that there was a marginal 1.4% increase in patient trust after the disclosure of the incentives. It was further reported that the increase was only among those who had a baseline high level of trust and not among those who had low trust at baseline. The authors also report that the disclosure had a greater impact on trust in the insurer rather than trust in the doctor (Hall, Dugan, Balkrishnan, & Bradley, 2002). Thompson et al., studied the impact of three interventions compared to control on the trust of the patients on doctors. When the patients enrolled in a HMO, three different types of induction visits were offered. The first was a physician-only visit, the second was a physician visit followed by a visit by a health educator and the last one was group visit of eight new members led by a physician and a health educator. In the group visit, the trust on physician was higher compared to control, but this was not seen in the other two groups. Overall trust in health plan was higher in the 'physician + health educator' and the group visit groups (Thompson, Gee, Larson, Kotz, & Northrop, 2001). A systematic review of interventions to build and promote trust in physicians and health plan revealed that there is insufficient evidence to show that there is any intervention that can promote trust in doctors (Mckinstry, Ashcroft, Car, Freeman, & Sheikh, 2008).

If it is true that interventions aimed at trust building do not actually build trust, what does this mean for measuring trust as a quality assessment strategy? Is trust a changing construct or is it a trait that a patient possesses? Are there trusting and non-trusting patients? These questions start arising and pose a challenge to using trust building as a quality improvement strategy. Empirical studies have shown that trust is only a state and not a trait. This means that trust can change based on context and circumstances. It cannot be a trait because there seems to be difference in the level of trust the same patients have on different physicians. There also seems to be no correlation between trust in physicians and overall trust in people. These indicate that the strategies used to build trust in the empirical studies described above were not the most appropriate ones. There is a need for more empirical work on trust building as a quality improvement exercise.

Using trust in physicians as a metric of quality of health care is potentially feasible. This is because trust can be measured in a valid and reliable manner using just the 12 items described in the trust in physician scale developed in this study. Not only is this a simple and valid tool, it is also easy to apply and monitor over time. There are three major applications of the trust in physical tool for assessment of healthcare quality. Firstly, it can be used in a cross-sectional manner to assess the level of trust at various points in time. Secondly, it can be followed up longitudinally to study the trends and patterns in levels of trust. Finally, it can also be used during specific stressful situations like an outbreak of an illness, or a major change in the government policy to study the possibility of mistrust, adoption or acceptance of the policy. In the era of Universal Health Coverage as a major Sustainable Development Goal, such metrics are very important and help to monitor progress towards these goals.

In summary, there is a broad scope for adoption of trust in physician as a quality metric and as a quality improvement strategy in health care. This is justified because trust in physicians is strongly related to trust in the health system and it feeds into the trust in health system in complex adaptive manner. Moreover, trust in physicians is a simple and easy to measure tool that can be used to understand the overall construct of quality of the health care through its interaction between processes and outcomes of the health care. There is a scope for use of the tool developed in this study as a metric of healthcare quality in developing healthcare settings with their unique limitations and uncertainties in delivery of health services. However, further studies are required to establish evidence for interventions to change trust in physicians and the health system.

References

Balkrishnan, R. (2003). Trust and satisfaction with physicians, insurers, and the medical profession. *Medical Care, 41*(9), 1058–1064.

Bonds, D. E., Camacho, F., Bell, R. A., Duren-Winfield, V. T., Anderson, R. T., & Goff, D. C. (2004). The association of patient trust and self-care among patients with diabetes mellitus. *BMC Family Practice, 7,* 3–9. https://doi.org/10.1186/1471-2296-5-26.

Donabedian, A. (1988). The quality of care: How can it be assessed? *JAMA, 260*(12), 1743–1748.

Hall, M. A., Dugan, E., Balkrishnan, R., & Bradley, D. (2002). How disclosing HMO physician incentives affects trust. *Health Affairs, 21*(2), 197–206.

Lee, Y., & Lin, J. L. (2009). The effects of trust in physician on self-efficacy, adherence and diabetes outcomes. *Social Science & Medicine, 68*(6), 1060–1068. https://doi.org/10.1016/j.socscimed.2008.12.033.

Mckinstry, B., Ashcroft, R., Car, J., Freeman, G. K., & Sheikh, A. (2008). Interventions for improving patients' trust in doctors and groups of doctors. *Cochrane Database of Systematic Reviews,* (3).

Mohseni, M., & Lindstrom, M. (2007). Social capital, trust in the health-care system and self-rated health: The role of access to health care in a population-based study. *Social Science & Medicine, 64,* 1373–1383. https://doi.org/10.1016/j.socscimed.2006.11.023.

Ohya, Y., Williams, H., Steptoe, A., Saito, H., Iikura, Y., Anderson, R., et al. (2001). Psychosocial factors and adherence to treatment advice in childhood atopic dermatitis. *Journal of Investigative Dermatology, 117*(4), 852–857.

Piette, J. D., Heisler, M., Krein, S., & Kerr, E. A. (2005). The role of patient-physician trust in moderating medication nonadherence due to cost pressures. *Archives of Internal Medicine, 165*(15), 1749–1755.

Thom, D. H. (2000). Training physicians to increase patient trust. *Journal of Evaluation in Clinical Practice, 6*(3), 245–253.

Thom, D. H., Hall, M. A., & Pawlson, L. G. (2004). Measuring patients' trust in physicians when assessing quality of care. *Health Affairs, 23*, 124–132. https://doi.org/10.1377/hlthaff.23.4.124.

Thompson, M., Gee, S., Larson, P., Kotz, K., & Northrop, L. (2001). Health and loyalty promotion visits for new enrollees: Results of a randomized controlled trial. *Patient Education and Counseling, 42*(1), 53–65.

Trachtenberg, F., Dugan, E., & Hall, M. A. (2005). How patients' trust relates to their involvement in medical care. *Journal of Family Practice, 54*(4).

Chapter 9
Building Trust Through Building Trustworthiness

Abstract Can trust be built? In the context of growing mistrust on physicians and healthcare systems among communities, what is the solution? These questions are difficult to answer. A sound analysis of the concept of trust leads to the understanding that trust building comes out of building trustworthy physicians and trustworthy health systems. To make trustworthy physicians, medical education reforms are essential. This chapter reflects on important measures to be adopted by the health system to address trustworthiness.

Keywords Trust building · Trustworthiness · Medical education

There is a growing discussion on trust deficit in health care globally. The trust deficit has been perceived not only in the physician–patient relationship, but also in the relationship between communities and health systems at large. Several factors have been reported to be associated with the deficit in patients' trust in their physicians. Increasing commercialization of health care has led to substantial conflicts of interests and doubts among patients whether the primary interest of their physicians is their welfare. Increasing technological applications in health has led to de-personalization of medical care and overreliance on medicines and gadgets, thus leading to trust deficits. Easy access to information through nearly universal availability of Internet access to all through mobile phones has led to access to unreliable sources of information, thus putting the medical opinion under doubt. The media coverage of medical scandals, research misconduct and cases of medical negligence, many times with exaggeration, has also eroded trust. Easy spread of rumours and misinformation through the powerful social media has also put the trust in the physician–patient relationship in jeopardy (Calnan, Rowe, & Hall, n.d.; Gopichandran, 2013). In this era of growing mistrust in the health system, what is the role of trust in the physician–patient relationship? Does it still matter? Is it even possible to build trust in the physicians? I will address some of these questions in this chapter and argue that the most important way to address trust deficit in the physician–patient relationship is to address the issue at the level of medical education. I will argue that building trustworthy physicians is the most sustainable and reliable solution to building and restoring trust in physicians.

V. Gopichandran, *Dynamics of Trust in Doctor-Patient Relationship in India*,
SpringerBriefs in Ethics, https://doi.org/10.1007/978-981-15-0346-7_9

With technology increasingly replacing the human aspect of medical care, and with the advent of artificial intelligence in medicine, the role of trust in physicians is likely to change. However, trust is a fundamental attribute of any human social interaction, and if the social interaction of health care exists, trust will continue to remain an essential element in it. It may transform from trust in physicians to trust in Watson (an artificial intelligence model in health developed by IBM), but trust will continue to influence the social interaction.

Experiments that have attempted to build trust in physicians have not demonstrated success as seen in the previous chapters. This is because they have approached trusting behaviours—demonstration of empathy, improvement of communication styles, etc.—none of which address the core aspect of trust, the intention of the physician. As seen earlier, the patient must believe that the intention of the physician is genuine for them to trust the physician. Many times, assessment of the intention of a physician or any human being for that matter is made at a subconscious level. Multiple verbal, non-verbal and social cues are processed at the subconscious level and lead to a feeling of trust. Though objective methods of assessment of trust have been developed, and one such scale is also presented in this book, trust is a result of a feeling or an emotion that arises in the mind of the person who trusts about the trustee based on this subconscious processing.

Trust building through manipulation of trusting behaviours of physicians sounds feasible but have not been successful. In this study, the previous chapters have described the various factors that seem to influence trust in the physicians. For example, shared identity, speaking the same language, providing comfort to the patient, knowing the patient personally and striking up a personally involved conversation, demonstrating kindness, empathy, etc., have been shown to influence trust. It is important to note that all of these can be potentially taught to physicians. Physicians can be taught communication skills, good bedside manners, empathy and the art of thorough history taking and clinical examination. However, these trusting behaviours do not always reflect the true intentions of the physicians. By providing the training in communication, physician behaviour, empathy and compassion, a physician can only be taught to exhibit these appropriate attributes in a clinical encounter. All these behaviours could be practised well by even a non-trustworthy physician who lacks integrity. Physicians could pretend to be kind, compassionate and could be brilliant communicators at the same time having ill intentions in their mind. This can be dangerous because the patients, who fall for these trusting behaviours, are left with no way to assess the true intentions of the physicians and hence face the risk of trusting the wrong person. Therefore, trust building through manipulation of trusting behaviours is a superficial process. This is probably the reason why trust building activities did not reflect as increase in trust in the physician.

If trust must be truly built and nourished, it must be a process of building trustworthiness. Building trustworthiness is a difficult and long-term process which must start at the level of medical education.

Trustworthiness is a virtue. Firstly, the medical student should understand the virtue, feel the importance of the virtue in the practice of medicine and should internalize the intrinsic value of the virtue. Every medical student should imbibe the attitude that they should be trustworthy physicians. The key attributes of trustworthiness are honesty, integrity, transparency, fidelity and altruism. All of these are important in a trusting physician–patient relationship. They all are attitudes of the physician which direct all their actions the ethically appropriate manner. Trustworthiness is not a matter of doing, but a matter of being.

Several models of medical education have been proposed, which take medical students close to the feeling emotions and lives of patients. These are likely to help students see suffering and pain from close quarters and thus develop an internal attitude of trustworthiness.

Community-based and community-oriented medical education is a simple strategy to take students to the community close to the patients and help them understand the needs of the people, their sufferings and reflect on their own role in such a society. In most medical schools in developing healthcare settings, medical education happens in the ivory towers of tertiary care hospitals. These are very far removed from actual communities. If medical education is taken to communities, close to people, then students are put shoulder to shoulder with the poor and needy. They see the actual realities of disease and suffering. This helps them develop an attitude towards patients, which promotes trustworthiness (Worley, 2002).

Humanities must be included and taught in the medical curriculum. The study of humanities helps medical students critically engage with the society, social structures, social norms, cultural beliefs and values. Thus, the students get an understanding of what it means to be sick and what is the role of the society in sickness. This understanding helps them place themselves in the larger scheme of things and inculcate an attitude of trustworthiness. When students see, perceive and understand the helplessness that often accompanies social determinants of health that the poor go through, they learn a lesson in humility and perceive the need to be trustworthy. The study of literature, poetry and art appreciation takes medical students close to human feelings and emotions. This undoes the systematic dehumanization of medicine that happens in medical college teaching (Ousager & Johannessen, 2010; Shapiro & Rucker, 2003).

One method that is very effective in trying to understand and empathize with patients is writing and reading reflective narratives of patients. Medical students should be encouraged to read reflective narratives of patients as well as write their own reflections on seeing patients and interacting with them. This helps students get in close tough with realities of medical practice (DasGupta & Charon, 2004).

Medical students can be systematically exposed to perform acts of altruism, honesty, fidelity and integrity. Organizing blood donation camps, making amends to someone to whom they have wronged and saying thank you to someone who has been helpful to them, all these help them experience acts of altruism, honesty, fidelity, transparency and help them feel good about these traits (Meili, Fuller, & Lydiate, 2011).

In addition to these methods, ethics and professionalism must be taught in medical schools. They should not be stand-alone modules as such modules tend to create an impression that they are disconnected from the routine practice of medicine. Medical ethics and professionalism should be taught as part of everyday clinical teaching. In olden days, ethics and professionalism are used to be learned through role modelling and observation of the senior practicing physicians. The same can now be done by specific instructions on ethics and professionalism by the clinicians during routine clinical teaching.

When medical education adopts these methods, it can significantly impact the attitudes of the medical students. It can influence the medical students to be trustworthy. Creating trustworthy physicians thus becomes the most important task in front of the medical education system. This must be carried out with urgency and diligence. Such trustworthy physicians will not just do the right things; they will be the right physicians.

Though the solution to trust deficit is in creation of trustworthy physicians, that does not mean there is no hope for the current generation of physicians and the community. Medical ethics, professionalism and trustworthiness training should be provided to practicing physicians. It must be more than just teaching communication skills and other neglected soft skills. It should involve adult learning techniques which are aimed at attitudinal change. There is much to be learned from evolutionary biology to influence attitudes of people. These techniques should be adopted to impart attitudinal change among practicing physicians, like continuing medical education programmes.

In summary, trust building must be approached as an activity of building trustworthiness. If we make the physicians trustworthy, people will automatically trust these physicians. The task is not easy. It is long drawn and difficult. But there seems to be no alternative to this method.

References

Calnan, M., Rowe, R., & Hall, C. (n.d.). Trust in health care.

DasGupta, S., & Charon, R. (2004). Personal illness narratives: Using reflective writing to teach empathy. *Academic Medicine, 79*(4), 351–356.

Gopichandran, V. (2013). Trust in healthcare: An evolving concept. *Indian Journal of Medical Ethics, 10*(2), 79–82. Retrieved from https://ezp.napier.ac.uk/login?url=http://search.ebscohost.com/login.aspx?direct=true&db=cmedm&AN=23697485&site=ehost-live.

Meili, R., Fuller, D., & Lydiate, J. (2011). Teaching social accountability by making the links: Qualitative evaluation of student experiences in a service-learning project. *Medical Teacher, 33*(8), 659–666.

Ousager, J., & Johannessen, H. (2010). Humanities in undergraduate medical education: A literature review. *Academic Medicine, 85*(6), 988–998.

Shapiro, J., & Rucker, L. (2003). Can poetry make better doctors? Teaching the humanities and arts to medical students and residents at the University of California, Irvine, College of Medicine. *Academic Medicine, 78*(10), 953–957.

Worley, P. (2002). Relationships: A new way to analyse community-based medical education? (Part one). *Education for Health, 15*(2), 117–128.

Chapter 10
Conclusion

Abstract This concluding chapter summarizes the strengths and limitations of this study on trust in physicians. There are three possible lines of enquiry that this research leads to—exploring trust in physicians as a quality indicator in health care, developing trust as a key ethical value in medicine and exploring medical education methods to improve trust in physicians through building trustworthiness.

Keywords Trust in physicians · Trustworthiness · Trust building · Quality in health care

This journey of trying to understand trust in the physician–patient relationship started from my work in the community in Tamil Nadu where I started working after graduating from my doctoral degree in community medicine. In the small village clinic, I was trying to implement evidence-based protocols and treatments for common conditions. One of the biggest challenges that we faced was trying to dissuade patients from demanding injections for all ailments and intravenous fluid infusions for tiredness and body pains. This practice was rampant in the private hospitals and clinics in the nearby town. We tried explaining, rationalizing, negotiating and sometimes even paternalistically rejecting their demands. But the problem was huge, and it only led to dissatisfied patients and erosion of trust in the clinic and is us as practicing physicians. We are a non-profit organization, and hence, our treatments were priced very low. We heard narratives from patients that they would rather go to the for-profit private hospital in the town, travelling more than 10 km than visit our clinic, but had no choice but to come to our clinic because of financial constraints. In the beginning, as a new practitioner in the area, this was disheartening for me. This and several other instances in which private physicians practising irrational and unethical medicine were trusted and respected more than us, who were trying to do the best for the patients, got me thinking about what makes people trust their physicians. Why are these few physicians in private practice in the town, out there just to make a quick buck, trusted more than us? Why was the narrative among the people, 'if only we had money, we would prefer to go to them rather than you' so prominent? What makes them trust their physician—the outcomes of treatment, the communication skills,

V. Gopichandran, *Dynamics of Trust in Doctor-Patient Relationship in India*, SpringerBriefs in Ethics, https://doi.org/10.1007/978-981-15-0346-7_10

the fulfilment of their demands, the low cost, the high cost? These questions were overwhelming, and this led to the exploration of the dimensions and determinants of trust.

What started as a qualitative exploration slowly took shape and grew into this Ph.D. project and the research questions and findings themselves iteratively led the research in the direction it took. On the way, I found some answers to the questions that were troubling me, but most important, I stumbled upon more questions and puzzles in the physician–patient relationship. At every stage of the long journey of this research, there was a mixed feeling. My most important understanding was that trust is a very complicated construct, especially in the physician–patient relationship. The way trust is defined and understood from the sociological perspective in the developed healthcare contexts is not applicable in developing healthcare settings like India. I found some unique dimensions, which can only be found in societies that respect a bit of paternalism in the physician. It can only be explained where the physician is expected to take charge and make the best decisions for the patients. The poverty, illiteracy, poor access to health care and other determinants which make illness a state of extreme vulnerability put people in these societies under high levels of stress, and therefore, they expect a level of paternalism in their physicians. However, while this was the reality in the areas where I researched the idea of trust, mostly among vulnerable populations, I also had my own friends and relatives who had a different concept of trust in their physicians. People living in the upper strata of the society, with higher levels of education, socio-economic status and healthcare access, placed their trust in the physicians based on more objective criteria. It was confusing, as I had embarked on finding out what trust means in a developing healthcare context, and what I landed up finding was that there cannot be a unified way of understanding trust as a social construct across socio-culturally unique communities. I realized that even the definition of trust which I was proposing from my study is not one that can be generalized to all of India or not even all of Tamil Nadu. I was just contributing to the large theory of trust in physicians in which my findings are a few bricks in the edifice.

This study gave me the humility to understand that one does not have final answers for serious and difficult social questions. One just expands the understanding of the social issue by broadening the scope of the problem. In this case, I do not think that I have a comprehensive understanding of trust in physicians. I only have a piece of the jigsaw puzzle, which happens to fit into the larger theory of trust in physicians. More such research will deepen the understanding of trust in physicians as a social construct.

There are several important limitations of this study, which warrant a description in this chapter. Firstly, I would like to reflect on my sample and the participants in my study. I would like to describe how this could limit the generalizability of my findings. Across the various stages of this research, I have focused mostly on people from marginalized communities. In the qualitative component of the study, the participants were people living in a backward district of Tamil Nadu, deprived of proper livelihood, healthcare access and experiencing poverty. I also included migrant construction labourers, who again formed a marginalized community within the capital city of Chennai. Therefore, their perspectives reflect key aspects of marginalization and do not represent the opinions of the larger community in Tamil Nadu. These findings shaped the quantitative survey. However, they did seem to resonate with the people who responded to the quantitative survey. But the quantitative survey did not have the option of exploring other dimensions of trust in those settings. Therefore, I am a bit concerned about the generalizability of the findings of the study to a larger developing healthcare context.

This study has looked at trust in physicians from an anthropological and sociological lens. Therefore, the discussions are all looking at trust in physicians as an essential element of the social interaction which is the physician–patient interaction. However, trust in physicians is more than that. There is an important perspective of trust in physicians from the moral and ethical lens. This has not been explored in detail in this study. There is a need to reflect on trust in physicians as an ethical principle. What does it mean to have trust as a normative ethical principle? If trust is indeed a moral value in medicine, what does it mean to build trust in the physician–patient relationship? What are the positives and negatives of articulating trust as an ethical value? Trust can be a double-edged sword—trusting less can deter utilization of services and at the same time trusting more can make people susceptible to exploitation. The ability to articulate trust as a moral principle lies in the balance that can be struck between these two extremes. How can this be done? This is an area that requires research but was not covered in this study.

Another limitation of this study was that it largely focused on the perspectives of trust in physicians from the patient and community point of view. This is only one side of the picture. There is a need for a detailed exploration of how trust in physicians is viewed from the providers' perspective.

As a result of this research, I developed a course for medical students on imparting important skills to them relating to communication, professionalism, ethics, cultural and social sensitivity. I used a mix of adult learning techniques in these courses including socio-drama method. In the socio-drama, very similar to the role-play method, the students are given scenarios which they must act out. They are asked to stage the whole scenario as a drama, with all background settings, properties and realistic settings of other people in the scene. This gives them a realistic experience of enacting the scenario. Their drama is video-recorded and played back to them. Then, the whole class reflects on the various aspects of the drama. This is a highly experiential learning, and it leaves a strong impact on the students. I learned a lot from this experience. I had started doing these trainings among students who had completed their MBBS course and were in their internship. I realized that the internship is a late point in time for training medical students on these skills. It is important to include this training much earlier in the time of medical education, maybe even when they enter medical school.

The current socio-political scenario in India, with the year 2019, seeing a major protest by physicians against acts of violence by patients and their relatives against doctors who are in duty, puts trust in physicians as a matter of great importance. There are increasing reports of violence against physicians in duty from across the world. These reports are notable in countries with developing healthcare settings like China and India. This points to one important issue. As much as trust is difficult to build, it is easy to lose. One careless act of omission or one inappropriate word or gesture at a vulnerable time is all that is required to erode trust and flare up rage. The said incident that led to protests condemning violence against physicians in India was due to the death of an elderly man because of acute heart attack in a government hospital in Kolkata. The reason for the violence was reported to be poor communication about the death by the junior doctor on duty in the hospital at that time, for which he was attacked and suffered a serious head injury. There are also several reports of killing of physicians and healthcare providers from China due to mishaps in the hospital during treatment of patients. These events are symptoms of erosion of trust in physicians. There is an urgent need to remedy this situation globally.

We recently organized an open house discussion in our institution for medical students to discuss violence against physicians and strategies to solve the issue. One of the major points that emerged from the discussion was the need for medical professionals and the community to give each other the benefit of doubt and room for error. It was also the need for medical professionals and the community to develop tolerance and understanding of each other's strengths and limitations. The overcrowded hospitals, long working hours, difficult working conditions and other inherent limitations in a typical government hospital setting can push the physicians to their limits. The same situations and in addition the vulnerability due to illness can push patients and their relatives over the board. It is under such circumstances that trust breaks and violence erupts. Maturity in thinking and understanding of one another seems to the solution. Reducing unrealistic expectations of the community from physicians through reasonable, realistic and honest campaigns is required.

This study has opened a few lines of enquiry into the area of trust in physician—patient relationship. All these areas of enquiry must be closely followed up in future research. I see three streams of work emerging from this research. One is research on using trust in physicians as a quality metric in health care. The feasibility of using trust in physicians as a quality measure and trust building as a quality improvement strategy needs to be explored in detail. This research must be informed by sociological, anthropological and health service principles. The second line of research is philosophical research to understand trust as a normative ethical principle and outlining the various aspects of trustworthiness as an inherent value in medicine. This line of enquiry should be guided by the discipline of philosophy and ethics. The third potential line of enquiry is to understand the ways to promote trustworthiness among medical students and physicians. What strategies work best? How can we promote trustworthiness in physicians? Is it even possible to evaluate trustworthiness? Such research questions need to be answered.

In conclusion, trust in physicians is an important and emerging area of clinical interest, social relevance and academic scholarship. There is a need for mainstreaming the discourse on trust in physicians in order to improve the social contract of the physician–patient relationship.

MIX
Papier aus verantwortungsvollen Quellen
Paper from responsible sources
FSC® C105338

If you have any concerns about our products,
you can contact us on
ProductSafety@springernature.com

In case Publisher is established outside the EU,
the EU authorized representative is:
Springer Nature Customer Service Center GmbH
Europaplatz 3, 69115 Heidelberg, Germany

Printed by Libri Plureos GmbH
in Hamburg, Germany